Sustainable Living at the Cer
Alternative Technology

This book presents a detailed exploration into the Centre for Alternative Technology (CAT), an enterprise concerned with finding and communicating sustainable ways of living, established in Wales in 1973. Playing a central role in the global green network, this study examines CAT's history and context for creation, its development over time and its wider influence in the progression of green ideas at the local, national and international levels.

Based on original archival and ethnographic research, this book provides the first in-depth analysis of CAT and uses the case study to explore wider issues of sustainability and environmental communication. It situates the Centre within current environmental and political discourse and emphasises the relevance and reach of CAT's practical solutions and creative educational programme. These practical solutions to the destruction of the environment of human activity are increasingly vital in today's context of climate change, loss of biodiversity and rising levels of pollution. It debates the spectrum of attitudes between environmentalism and ecologism evident at CAT and in broader conversations surrounding sustainability.

Woven throughout the text, the author makes clear what we can learn from CAT's almost 50 years of experiments and experiences, from his first-hand account of working at the site. This will be a fascinating and revealing read for academics, researchers, students and practitioners interested in all aspects of sustainability and environmental issues.

Stephen Jacobs is Senior Lecturer in Media, Religion and Culture at the University of Wolverhampton, UK. His recent publications include *The Art of Living: Foundation: Spirituality and Wellbeing in the Global Context* – a multi-sited ethnographic study of an international meditation organisation. He has had an interest in environmental issues and sustainability since working at the Centre for Alternative Technology in the 1980s. He was awarded a Leverhulme Research Fellowship for this project.

Routledge Studies in Sustainability

For more information on this series, please visit: www.routledge.com/
Routledge-Studies-in-Sustainability/book-series/RSSTY

Sustainable Living at the Centre for Alternative Technology

Radical Ideas and Practical Solutions

Stephen Jacobs

Routledge
Taylor & Francis Group

LONDON AND NEW YORK

earthscan
from Routledge

First published 2023
by Routledge
4 Park Square, Milton Park, Abingdon, Oxon OX14 4RN

and by Routledge
605 Third Avenue, New York, NY 10158

Routledge is an imprint of the Taylor & Francis Group, an informa business

© 2023 Stephen Jacobs

British Library Cataloguing-in-Publication Data
A catalogue record for this book is available from the British Library

Library of Congress Cataloging-in-Publication Data
A catalog record has been requested for this book

ISBN: 978-1-032-07562-4 (hbk)
ISBN: 978-1-032-07563-1 (pbk)
ISBN: 978-1-003-20770-2 (ebk)

DOI: 10.4324/9781003207702

Typeset in Bembo
by Newgen Publishing UK

Contents

Acknowledgements

This project would never have been realised without the generous support of the Leverhulme Trust. This research would not have been possible without the cooperation and support of everybody at the Centre for Alternative Technology (CAT). Many people from CAT, past and present, not only were prepared to give me their time, but were also very open about their experiences. In particular, I must thank Paul Allen and Peter Harper, who always very promptly and patiently responded to my many emails during the course of my research. Thanks must also go to Ruth Stevenson who was also incredibly supportive and helpful throughout this project. Thank you Anne Marie Carty for all the incredibly useful and stimulating conversations that we had at the beginning of my project. Thanks must also go to Allan Shepherd, who was the main instigator of the oral history project. The recordings and archive of CAT's history provide fascinating glimpses into CAT's past, which were invaluable to my research. Everyone at the National Library of Wales was also incredibly helpful in assisting me access the archives and finding all the documents that I needed. Also, I must thank my good friend and colleague Alan Apperley, who very kindly gave up most of his Easter holiday to proofread the manuscript, which now reads much better. Finally, this project would never have seen the light of day without the constant and unwavering support of my wife Judith.

Introduction

The Context: Fear Won't Do It!

There has been a dramatic increase in extreme weather events across the world: more frequent cyclones, unseasonable temperatures and an increase in droughts. The media is replete with images and narratives of devastating fires, catastrophic floods, disappearing Antarctic ice and the threat of rising sea levels as well as loss of biodiversity and polluted seas. Greta Thunberg, the young Swedish environmental activist, challenges world leaders to do more to address the climate emergency. Protest groups such as Extinction Rebellion have taken to the streets proclaiming there is no Planet B. Politicians across the globe at local, national and global levels have declared that humanity and planet earth face a climate emergency.

The UN Intergovernmental Panel on Climate Change (IPCC) in its latest report unequivocally states that there is a direct link between human activity, climate change and the extreme weather events witnessed this year across the globe. The report states:

> Human-induced climate change is already affecting many weather and climate extremes in every region across the globe. Evidence of observed changes in extremes such as heatwaves, heavy precipitation, droughts, and tropical cyclones, and, in particular, their attribution to human influence, has strengthened since AR5.
>
> (IPCC 2021)

This latest Report from the IPCC is not ambiguous. 'Climate change *is* "human induced" and *is* already affecting many weather and climate extremes'. The Report is replete with phrases such as 'high confidence', 'very likely' and 'virtually certain'. There are far fewer hedging phrases such as 'there are many uncertainties' and 'our judgement is that' found in the first IPCC Report on Climate Change published in 1990. This expression of certainty gives a sense of urgency as climate change is not simply something that will happen sometime in the future but is happening now. It is not something that is simply affecting

DOI: 10.4324/9781003207702-1

far-flung populations, distant coastal communities or sub-Saharan Africa – it affects everyone across the globe in one way or another.

While climate change deniers still exist and there are still a few who do not accept that this looming environmental catastrophe is caused by human activity, the summer of 2021 in many ways marks a paradigm shift in the discourse about climate change. Phrases such as 'tipping point', 'code red for humanity' and 'runaway effect' are increasingly used not only by climate scientists and IPCC but are also becoming ever more prevalent in political rhetoric and in the media. Climate change can no longer be dismissed as the exaggerated projections of deluded doom-mongers and mavericks when the German Chancellor Angela Merkel stated after visiting the areas devastated by floods in July 2021, 'We have to up the pace in the fight against climate change' (cited in Oltermann 2021). However, to date, the political, social and cultural responses have been totally inadequate.

There is a burgeoning literature on why there has been a totally inadequate response to this climate emergency. While there are numerous theories – psychological, social, political, economic – as to why the response to the climate emergency has been so perfunctory, I initially flag two interconnected reasons for the current inertia: first, the scale of the problem, and second, the dearth of positive discourses. The apocalyptic scale of the climate emergency is so overwhelming that humanity seems to be like a rabbit paralysed by the headlights of an unstoppable truck hurtling towards disaster. Saffron O'Neill and Sophie Nicholson-Cole (2009) argue that while spectacular images of the destructive force of weather extremes or starving polar bears and apocalyptic narratives of tipping points 'may well act as an initial hook for people's attention and concern, they clearly do not motivate a sense of personal engagement with the issue [of climate change]' (O'Neill and Nicholson-Cole 2009: 375). It is of course critically important that people actively engage with the threat of climate change, but as O'Neill and Nicholson Cole point out, 'fear won't do it'. However, in public discourses about the climate emergency, there is a paucity of positive proposals as to how to address this looming disaster.

John Urry also suggests that 'fear is not the best way to induce low carbon'. He poses the rhetorical question 'how could low carbon lives be innovated, generalised as a practical, desirable and fashionable set of alternatives?' (Urry 2011: 122). In this book, I argue that possible answer to Urry's question can be found at the Centre for Alternative Technology (henceforth CAT) located in rural mid-Wales. CAT was founded in 1973, and throughout its history, its focus has always been on finding practical solutions to environmental challenges. My argument is not so much whether the actual solutions *per se* proposed by CAT are effective, but that this positive solution-based discourse can inspire people to actively engage with environmental issues. This positive solution-based approach has the potential to not only mitigate the actual challenges that humanity and the planet face, but also address the slough of despondency that can be created by the constant bombardment of apocalyptic images and narratives. The deep ecologist Freya Mathews (1995: 125) observes: 'we cannot hope to change Western civilization, in the ways that we as conservationists dream of, if we constantly threaten it with the wholesale ecocidal consequences of its aggressive activities'.

Practical Utopianism

I call this positive solution-based approach to environmental destruction practical utopianism. The concept of a practical utopia is indicated by the sub-title *Radical Ideas and Practical Solutions*. Radical ideas refer to the utopian aspect of the project, and practical solutions indicate the pragmatic elements of CAT. The concept of practical utopia is also suggested by the term 'alternative technology' (AT) itself – alternative signifying the utopian aspect of the project, while technology denotes the practical aspects of the CAT project. Before looking more closely at what I mean by a practical utopia, I will briefly indicate two generally accepted perceptions of utopias that can be called the fictional and the social engineering models. I suggest these models are misrepresentations of utopianism, and drawing on the work of commentators like Ruth Levitas (2011, 2013) and Lucy Sargisson (2000, 2012), I suggest a third conceptualisation of utopianism that I call the remedial model. This remedial model views utopianism as an ongoing process rather than a clearly defined blueprint of the good society. CAT's practical utopianism can be thought of as a subset of the remedial model of utopia.

The fictional model suggests that utopia is an unrealisable fantasy. It only belongs in the realm of the imagination and manifests in the pages of speculative fiction. It is clear that Thomas More who coined the term 'utopia', in his novel of the same name, did not intend his vision to be a blueprint for society. As Ruth Levitas points out, the term coined by More is a joke and 'contains a deliberate ambiguity: is this eutopia, the good place, or outopia, no place – and are these necessarily the same thing?'(Levitas 2011: 2) Perhaps, the best way to interpret More's deliberate ambiguity is to suggest that the society of Utopia that he describes is not only a good place that does not exist, but also a good place that cannot exist. There are clearly aspects of the society that More describes in his pages that are not intended to be reproduced in any society. Perhaps, even more pertinent here is Ernest Callenbach's 1975 novel *Ecotopia*, in which he envisages a society on the Pacific Coast, which has declared independence from the United States of America (USA) and has adopted a more environmentally friendly lifestyle. However, the ecological society described in this novel is also clearly not intended to be realised.

The social engineering model of utopia indicates projects that are attempted in the real world, rather than simply described in the pages of speculative fiction. However, enacting utopian projects in the world almost inevitably becomes dystopian. The evidence seems indisputable. Some of the utopian projects of the 20th century, such as Maoism, caused untold suffering. There are three reasons why any attempt to implement a particular utopian dream in the world are doomed to failure. First, these utopian projects tend to be rigid and formulaic – there is only one model for 'the good society'. These utopias fail to acknowledge that there might be other models of a good society. Second, these models tend to be imposed from above. Finally, while the intent to improve life for all might be genuine, rather than a cynical political ploy, societies are highly complex and there will always be unintended consequences that are often detrimental to the population.

The remedial model indicates that utopianism is neither a fantasy nor a form of social engineering, but an experimental space. Levitas (2011: 8) suggests, utopian thought is 'a desire for a better way of being and living'. However, 'Utopia is not just a dream to be enjoyed but, a vision to be pursued' (Levitas 2011: 1). In this sense, utopia can be thought of as a journey and not a destination. A desire for 'a better way of being and living' necessarily involves a critique of various aspects of contemporary society. In Lucy Sargisson's terms, utopianism is transgressive as it challenges the status quo and commonly held assumptions. Utopias are experimental conceptual spaces not only for diagnosing society's ills but also for suggesting potential remedies. These remedies are not fixed absolutes, but explorations of new possibilities. Lucy Sargisson (2012: 7), drawing on the ambiguity of More's term, suggests that utopia 'poses a conflict between desire and realization … it lies always over the horizon'.

How we pursue a better way of being, in this instance a more environmentally sustainable way of being and living, entails a practical aspect. Here I draw on aspects of what Erik Olin Wright (2010) calls 'real utopias'. Wright's concept of a real utopia is a form of remedial utopia as it involves critique and suggests remedies, but more directly informs the 'tasks of navigating a world of imperfect conditions' (Wright 2010:6). I briefly summarise Wright's ideas before discussing why I describe CAT as being a practical utopia rather than using Wright's terminology.

Wright begins by suggesting that:

> The idea of 'real utopias' embraces the tension between dreams and practice. It is grounded in the belief that what is pragmatically possible is not fixed independently of our imaginations, but is itself shaped by our visions.
>
> (Wright 2010: 6)

Since its very beginning, CAT has been shaped by the vision of a more sustainable way of being. It has always tried to find practical means to achieve those ends.

Wright (2010: 8) suggests that real utopias have three tasks – 'diagnosis and critique; formulating alternatives; and elaborating strategies of transformation'. The starting point, Wright (2010:11) argues, is to identify 'the ways that social institutions and social structures systematically pose harms on people'. The second step is to 'develop a coherent, credible theory of alternatives' (Wright 2010:20). These alternatives must be desirable, viable and achievable. Wright (2010:22) suggests that there are many viable alternatives for mitigating social harms, and that 'the exploration of viable alternatives brackets the question of their practical achievability'. However, 'viable alternatives are more likely to become achievable alternatives if they are well thought out' (Wright 2010: 23). Developing achievable alternatives is contingent upon:

> The extent to which it is possible to formulate coherent and compelling strategies which both help create the conditions for implementing alternatives in the future and have the potential to mobilize the necessary social forces to support the alternative when those conditions occur.
>
> (Wright 2010: 25)

Finally, strategies of transformation 'tell us how to get from here to there' (Wright 2010: 26). The strategies of transformation identify the methods by which social harms may be remedied. This is neither a fantasy nor a form of social engineering, but Wright clearly views real utopias as a process of transformation and not as seeking a state of final perfection. The threats to the environment have been clearly identified, and there is a desperate need to explore viable alternatives to develop achievable ways to ensure that life is sustainable. In this book, I argue that CAT's nearly five decades of exploring viable alternatives can and have helped elucidate achievable options to address the climate emergency.

Wright briefly mentions environmentalism. He suggests that 'capitalism significantly contributes to environmental problems' (Wright 2010: 69). He identifies two reasons for capitalism having a detrimental effect on the environment. Firstly, capitalism's prime focus is on maximising profits, which often takes precedence over environmental concerns. Second, 'nonrenewable natural resources are systematically under-priced in the market since their value to people is not registered in the dynamics of supply and demand in the present' (Wright 2010: 69). In other words, Wright's argument is primarily a critique of capitalism and environmental harms are a direct result of the capitalist system. While there is an implicit criticism of the global capital marketplace in CAT's rhetoric, this is secondary to the urgency of directly addressing environmental issues. While Wright refers to a couple of concrete examples of real utopias, such as participatory city budgeting, *Wikipedia* and the Mondragon worker-owned cooperatives, his work remains primarily speculative and theoretical. CAT, on the other hand, has always actively explored and experimented with viable alternatives in the world.

CAT is utopian because it envisages a better more sustainable way of being. The focus of CAT experiment is to find practical solutions to the environmental issues faced by humanity and the planet. Consequently, I prefer the term 'practical utopia' to describe CAT. Another reason why I prefer the term 'practical' over 'real' is that 'real' suggests an ontology. The concept of a 'real utopia' suggests that utopian projects in the world, like CAT, have an ontological status different from a work of utopian fiction. Nonetheless, both 'real' and 'fictive' utopias can be considered as experimental spaces for imagining better ways of being. Furthermore, CAT has always been pragmatic. CAT has, as I will show in delineating its history, always made changes when circumstances required.

Nonetheless, Wright's model is useful. CAT has clearly diagnosed an issue, that much human activity, particularly post-Industrial Revolution, is inherently environmentally destructive. Implicitly, and often explicitly, this diagnosis critiques the current state of affairs – often expressed as business as usual – as not sustainable. CAT has explored a number of viable alternatives to 'business as usual'. Some of these viable alternatives have proved not to be effective for various reasons, and so CAT has not been afraid to adapt and change in the light of new evidence or a change in context. CAT's exploratory approach has always been driven by a desire to suggest *achievable* alternatives to 'business as usual'.

This sense of achievability is articulated in terms of CAT's claim that we already possess the means to address the environmental issues faced by humanity and the planet. Finally, CAT has always elaborated practical and viable strategies to enable humanity to live more lightly on this planet. Tim, who worked at CAT from 1982 to the mid-1990s, encapsulates the idea of a practical utopia when he suggests:

> CAT was very down to the ground and very in the clouds at the same time and successfully spanning both.
>
> (Tim, interview February 2021)

Remote, Yet Not Too Far!

CAT is tucked away in a relatively remote corner of rural mid-Wales and is not as widely known as Friends of the Earth (FoE) or Greenpeace. Nonetheless, I argue that CAT has and continues to play a significant role in the environmental movement, and its influence is much greater than is immediately apparent. I deliberately used the term 'remote'. As Zoe Gardner points out, remoteness was an integral concept in the philosophy of early AT. The geographical remoteness and rural location reflected the idea that AT was inherently critical of mainstream modern life epitomised by the city (Gardner 2008: 296). However, this repudiation of prevailing mores has created a double bind. On the one hand, the ideas that CAT promulgates may seem remote; on the other hand, CAT has an imperative to inform and convince a wider public of the need to address the environmental crises. Consequently, I consider whether the ideas and values that underpin CAT specifically, and the environmental movement more generally, are regarded as inaccessible or irrelevant, and what the strategies are for overcoming that perceived cognitive distance. In other words, one of the central concerns of this book is to consider the relevance of CAT, and how this can inform an understanding of the challenges of prompting individuals, organisations and governments to actively engage with the anthropogenic causes of the climate emergency and other environmental crises.

It could be suggested that its remote geographical location has been detrimental to CAT's effectiveness. CAT is located in the Dyfi Valley, which is a rural area with a scattering of small villages and sheep farms. The nearby market town of Machynlleth has a population of under 2,500. Slate quarrying used to be a major employer in the region, though the vast majority of quarries are now closed. Indeed, CAT is itself located in the disused Llwyngwern Quarry, and consequently, the site itself is often affectionately referred to as The Quarry. Surely, it would have been better to have been located nearer to some large urban conurbation with easier access routes. The current CEO speculates that:

> In some ways, it makes complete sense to up sticks and relocate to a business park off the M6 with access to really good centres of population and really good communications links – but it wouldn't be CAT.
>
> (Peter, interview April 2021)

The Quarry has a very particular *genius loci*, which has contributed to CAT's ethos, and almost all the people who have been or are still involved in CAT make some reference to the importance of the site:

> I think there's something about the actual physical place that has its own soul somehow. There's something wonderful about that place. And it's partly the actual physical place that many people fall in love with.
>
> (Sally, interview May 2021)

A further aspect of the place itself is the historical narrative. There are two aspects to this historical narrative that contribute to the importance of the site itself: reclaiming the landscape and the people. Llwyngwern Quarry, despite its rural setting, might be considered as an industrial wasteland, similar in many ways to the brownfield sites of the failed industries more commonly found in urban settings. Slate has been quarried in Wales for local housing for centuries. However, the demand for slate increased dramatically in the nineteenth century with the twin processes of urbanisation and industrialisation. Slate quarrying is notorious for the amount of waste that it produces. A 1946 report on the Welsh slate industry suggested that every ton of usable slate produced 20 tons of rubble, and this, as C.T. Crompton observes, has blighted the landscape (Crompton 1967: 291–2). A photograph of Lwyngwern Quarry taken in 1973, and now displayed on-site, shows an almost barren landscape devastated by the slate waste. However, one of the earliest projects at CAT was to build up the soil and create a garden. The Quarry is now a lush and vibrant oasis. This transformation of the landscape is represented as reclaiming the industrial wasteland and consistent with the ideals of sustainability. Allan Shepherd observes that 'built on a disused slate quarry' became a prevalent phrase in descriptions of CAT and contributed to 'a phoenix mythology. If a vision for the future could be created on the most earth-torn of landscapes, it was possible to turn around any desperate situation with imagination' (Shepherd 2015a: 43).

The second aspect of the historical narrative suggests that the commitment and hard work of the many people who have passed through the Quarry over the decades give CAT tangibility and weight. It is the history of all the people who have contributed to CAT in so many ways that provides the project with a credibility that purely online organisations do not have.

You could suggest that CAT is an obscure group that has never had the impact that other environmental organisations, such as FoE, have had. It is just a small group of 'crazy idealists' to utilise the title of the booklet that celebrated CAT's twentieth anniversary. These idealists, although highly motivated and committed, have been accused of being out of touch with most people. Other than those already converted to the cause, some people who have encountered CAT have simply perceived it as a rather curious anomaly. On a visit by Prince Charles in 1978, one of the accompanying dignitaries was overheard to say, 'There is nothing for us here'.[1] Some in the local Welsh community, and others, have referred to CAT as 'the hippies on the hill'. However, I argue that the assessment that CAT has nothing to offer the wider world is fundamentally mistaken and that we can learn a great deal about the challenges of living sustainability from the almost five decades of CAT's experiences.

Furthermore, to simply dismiss CAT as hippies is also a misunderstanding. While some countercultural ideas informed early environmental thinking (see Kirk 2007 and Zelko 2013), neither CAT nor the broader environmental movement can be considered, despite some outsiders' perceptions, as countercultural. The counterculture is famously defined by Theodore Roszak as 'a culture so radically disaffiliated from the mainstream assumptions of society that it scarcely looks to many as a culture at all, but takes on the alarming appearance of a barbaric intrusion'[2] (Roszak 1995: 42). Most, albeit not all, environmentalists tend to challenge the mainstream presuppositions – particularly the dominant neoliberal interconnected discourses of the logic of the global marketplace, that economic growth is inherently beneficial for all, and that the acquisition of material goods is a marker of success. However, the challenge for CAT, and advocates of a more sustainable lifestyle, is to oppose some of these taken for granted assumptions of modern society without appearing to be 'a barbaric intrusion'.

The title of the CAT's 20th anniversary booklet *Crazy Idealists!* was rather tongue in cheek – note the exclamation mark in the title. Paul, who has worked at CAT since 1988 with, as he observes, 'various different hats on', suggested that:

> Crazy idealists was used by mainstream media to describe CAT in the early days.
>
> (Paul, interview January 2021)

It is also of note that a later edition of the CAT story bought out for the 25th anniversary slightly changed the title to *Crazy Idealists?* The ironic use of the exclamation mark and the later question mark was intended to challenge the perception that CAT is a romantic and unrealistic project. The implication is that CAT is neither crazy nor idealistic but is eminently sane and practical. CAT is a prime example of what Andrew. G. Kirk has termed 'environmental pragmatism' (Kirk 2007: 143) as it aspires to find realistic, practical solutions to the challenges of environmental crises. From the very beginning to the current day, CAT's ethos has always been positive – briefly alluding to the environmental challenges, but primarily emphasising how these challenges can be resolved. Paul Allen recalled that he had been an active member of FoE when he first visited CAT in 1979 with a friend:

> When I arrived at this place in Wales where people were not protesting against what they thought was wrong. They were working for what they thought the solutions were. That seemed very different somehow. And so instead of slapping a notice on the gate of Sellafield saying close it, people were working in developing alternatives to that, which is absolutely essential. If you're going to say, no, I don't want this. People are bound to say, well, what shall we have then?
>
> (Paul Allen, Oral History)

The CAT project goes further than simply trying to find sustainable solutions to environmental crises. Since the very early days, education and communication have been an essential aspect of the project. In 1975, CAT opened as a visitor centre. At its peak in the mid-1990s, there were over 90,000 visitors to the site per annum although visitor numbers did start to decline in the early 2000s. Nonetheless, the Quarry is still a popular destination. Overall, there is a spectrum of visitors to CAT with a diverse range of knowledge, presuppositions and expectations. The heterogeneity of visitors is reflected in the very different comments on TripAdvisor, which range from 'a massive opportunity to educate and inspire missed at almost every level' (Muffinn 2: TripAdvisor August 2021) to 'So inspiring. For anyone interested in how to live better and live right by our planet this is a must visit' (Mojo: TripAdvisor August 2021).[3] This diversity poses a challenge to CAT that is analogous to the wider environmental movement's imperative to appeal to as wide an audience as possible. By 2015, Allan Shepherd estimated that more than 3 million people have visited CAT, contacted its information service or attended a short course (Shepherd 2015b: 2). CAT has had an impact on many of these visitors, schoolchildren and students, which not only has influenced individual lives, but also has contributed to CAT's international reputation.

It is also possible to suggest that CAT is anachronistic as many of the environmental issues that CAT was concerned with in the early decades of its existence have largely been resolved. For example, recycling and composting are now commonplace, whereas in the 1970s and 80s, these were marginal practices at best. Renewable energy, which in the 1970s was largely regarded as ineffective and irrelevant, is now mainstream. People may well have photovoltaic panels on their roof, and wind turbines are now a common sight. So one could say 'job done', and conclude that CAT no longer has a significant role to play. However, in the nearly 50 years of its existence, CAT has survived where other similar experiments have failed. There are several reasons for CAT's survival: It has been highly adaptable; it has inspired an incredible level of dedication to the organisation by those who have lived and worked there; and it has retained its independence from both industry and government.

For example, The Earth Centre in South Yorkshire was intended to be a visitor and educational centre about sustainability. It was opened in 1999 at a cost of £55 million, substantially more than the start-up money for CAT. The Earth Centre was primarily funded by the Millennium Commission, European Commission and English Partnerships. The Executive Director described the project:

> With a further dramatic increase in world population by the middle of the next century and compound economic growth, global society may need to be at least 10 times more efficient in our use of resources than we are today. Earth Centre is about facing up to that challenge and trying to inspire people and involve them in what is unquestionably the key issue of our age. In a few years, as the project matures and as new features are added, we are confident that we can make a bold and unique contribution to the advancement of sustainability.
>
> (Smales 1999)

The Earth Centre was located on a disused colliery, so had many parallels with CAT as part of its narrative was about reclaiming a derelict industrial land-scape in a sustainable way. However, the Earth Centre went bankrupt in 2004 as it failed to attract sufficient visitors and had to close its gates. Like many Millennium projects, the Earth Centre was overly ambitious. According to Nonie Niesewand (1998), it was hoping to attract 500,000 visitors in its first year. While, the Earth Centre was not as remote as CAT, with the urban centres of Doncaster, Sheffield and Leeds easily commutable by public transport or car, this aspiration was probably never realisable.[4]

The question of why CAT has survived and how it has retained relevance is a central feature of this book. This resilience is, at least in part, due to CAT's ability to adapt to changing circumstances. I suggest that CAT has mostly been ahead of the curve in terms of responses to environmental issues. In its very early days, CAT was well ahead of the game, and perhaps a little too alien for many except as an exotic curiosity. In the mid-1990s, it was possibly a little behind the environmental discourses and possibly slightly anachronistic. However, for much of its existence, CAT has caught the wave just right – being just as far enough ahead to be innovative, but not too far in advance to appear to be irrelevant.

This case study of CAT casts light on the interrelationship between radical ideas and practical solutions in discourses of sustainability. Consequently, whilst there is a historical dimension to this research, I utilise this longitudinal narrative to elucidate the challenges to sustainability posed in today's context of the climate emergency, loss of biodiversity and other threats to the environ-ment. Overall, I argue that whilst CAT might initially seem to be an obscure small group of idealistic mavericks secluded in a remote corner of rural UK, it has always been and continues to be a significant player in the environmental movement. CAT has had a global reach that is perhaps not so obvious as large international groups such as Greenpeace or FoE and has not been as visible as Extinction Rebellion. People from all over the world have heard of CAT, visited the Quarry and been inspired by its dual ambition to identify practical solutions to the detrimental impact of human activity on the environment and communicate these solutions to the wider public. Some of CAT's projects have failed; some have worked for many years and then for various reasons ceased to be effective. Indeed, one of the mantras of CAT is that 'failure is the compost of success'. There is a great deal to learn from CAT's experiments with sustainability that is relevant to today's environmental crises. CAT is also the central node of a network of individuals and groups that have in some way been inspired by its quest for practical solutions. CAT has an invisible network similar to the mycelium of fungi, which has spawned other productive offshoots, which in turn produce others.

> CAT is like a mothership with satellites which are being able to take the message out to people who wouldn't reach it directly.
>
> (Sally, interview May 2021)

CAT's survival over five decades has also contributed to its credibility, and it is now widely accepted as an authority in sustainable living by those who would have been dismissive of it at the beginning. Environmentalism has come in from the cold, and this presents a whole range of different challenges. For CAT, this move from margins to mainstream has primarily been a challenge for the development of the visitor circuit. The vast majority of people are now aware of climate change and other environmental crises. It is no longer sufficient, if it ever was, to simply inform, but also as CAT's mission statement indicates, it is also necessary to inspire and enable people to engage with the current environmental crises.

Methodology

It will be obvious by now that I am sympathetic towards the CAT project. In other words, I have a particular standpoint in relationship to the research. It is now widely acknowledged that all researchers have a particular perspective on their areas of interest. Neutral objective accounts, or what Donna Haraway characterises as 'the god trick of seeing everything from nowhere' (Haraway 2004: 86), are neither possible nor desirable. My perspective is informed by two interrelated factors.

First, I am convinced that we need to urgently address the climate emergency, loss of biodiversity and other environmental harms caused by human activity. My ecosophy, which Arne Naess (1995: 8) defines as 'a philosophy of ecological harmony or equilibrium', is broadly in alignment with that of CAT. I believe that CAT's experiments with sustainability should be more widely known. We can learn a great deal from CAT's aspiration to identify sustainable solutions to the anthropogenic causes of environmental degradation and its mission to communicate its insights and persuade a wider public to take action. To date, there is almost no mention of CAT in the academic literature. There is a section on CAT in David Pepper's 1991 publication *Communes and the Green Vision*. However, this book is out of date and Pepper wrongly associates CAT, as many people have, with the counterculture. Adrian Smith produced a short paper for the European Consortium for Political Research in 2004 in which he compares CAT with the short-lived Biotechnical Research and Development (BRAD), which was coincidentally also founded in 1973 and located in Wales. BRAD was disbanded after three years because of a lack of resources and internal disagreements. Smith describes both CAT and BRAD as R&D labs for utopia. Smith (2004: 15) suggests 'renewable energy, which was utopian in an early social context, now has a significant strategic relevance for society'. While I agree that renewable energy has now become of strategic significance with the concomitant improvements in the technology, increasing economic viability and political acceptability, I do not agree, as I will elucidate, that CAT has ceased to be a utopian project.

Second, my standpoint is informed by the fact that I worked at CAT before embarking on an academic career. In 1984, I was living in London and wanting to leave, when I saw an advertisement for a position to run the vegetarian restaurant at CAT. I had never heard of the place before, and unlike most of the people at CAT, I had never really thought about environmental issues beyond having some vague ideas about saving whales. However, I had trained as a chef and was (and still am) a vegetarian, so I decided to apply for the post, more out of curiosity than anything else. To my surprise, I was offered the position of Restaurant Coordinator and moved to Wales. I lived on-site and ran the restaurant for the next two years. I felt very much at home there, stayed in the area and worked on and off for CAT until 2000. I also was a founder member of a housing cooperative, which was begun in 1986, when a small group of us decided that we wanted to live off-site, but still maintain a low environmental impact lifestyle.

My involvement with CAT raises the question of the distinction between insider and outsider accounts. In ethnographic research, it is it is increasingly recognised that making an unambiguous differentiation between insider and outsider is highly problematic. Stephen E. Gregg and George D. Chryssides observe that 'this is not a simple binary matter of being "inside" or "outside", but is a fragmented, contextualised and sometimes contradictory, set of dynamics which build up a bricolage of identity factors' (Gregg and Chryssides 2019: 21–2). One must, as Gregg and Chryssides argue, understand interactions between researchers and researched as relational, rather than some sort of fixed and inherent status. Therefore, this research and publication is informed by my relationship with CAT and the people who worked there. Consequently, my standpoint can be considered as an ex-insider. This position has proved both a help and a hindrance. As an ex-insider, my relationship with CAT and the people who work or used to work there is complex. Obviously, both I and CAT have changed, and this means that I have a very different relationship with CAT than I did when I worked there.

CAT has undergone major changes since I left. Some of these changes have been quite traumatic for both individuals and the organisation as a whole, and I will address this in more detail later in this book. Methodologically, my association with CAT in the 1980s proved to be an obstacle, rather than an aid to access. While I could clearly ask my personal contacts, many of whom remain good friends, to help with my research, initially I was not so successful in engaging with the current senior management at CAT. There are numerous reasons why individuals might not wish to participate in an academic research project – for example, they might not have time, they might not see the research project as being important or they might simply feel that they do not have anything to contribute. However, I was faced by an additional challenge in recruiting research participants, which I had failed to anticipate. This challenge was linked to the history of CAT after I had left, which, although I was aware of, I had not fully considered. I simply, and wrongly as it turned out, assumed that my past involvement with CAT would open all doors.

In the 1980s, when I was there, CAT was run on a cooperative basis, with everyone having the opportunity to have an input into the decision-making process. However, for a variety of complex reasons, the consensus decision-making procedures were abandoned and a more conventional and hierarchical management system was established in 2010. This transformation of the organisational structure caused a great deal of grief, and some felt that this was a loss if not a betrayal of an essential element of CAT. The current senior management were initially concerned, because of my previous history with CAT, that I might be critical of the new organisational structure and romanticise CAT's earlier more alternative and consensus ethos. However, I did eventually manage to persuade them that this would not be my approach, and I have been able to interview a number of the senior management team, some current members of general staff and a couple of the current Board of Trustees. The CEO at the time, in his interview indicated:

> As you know, when you first made contact, I was a little reticent because, what we don't need is just hagiographies of the pioneers and how wonderful it was then and how it has now gone down the pan because it's sold its soul.
>
> (Peter, interview April 2021)

This support of the current senior management and the Board of Trustees is obviously essential for my research. However, this also creates a different challenge, in terms of writing up my project. The restructuring of CAT in 2010 was primarily brought about by a very serious financial crisis, which not only led to a change in the organisational structure, but also meant that some people, including some of my friends, were either made redundant or had their roles dramatically changed. This inevitably caused a great deal of pain and sometimes anger. This is a sensitive issue and raises the issue of how this very personal and emotional period of CAT's history is to be represented.

It is a truism to assert that when one involves friends in an academic research project, you enter into a different type of relationship. Once again, it is necessary to clearly state my standpoint in regard to this emotional history. First, this traumatic period of CAT's history did not directly affect me. Second, it is a moot point whether or not CAT would have needed such a root and branch restructuring if it had not faced such a severe financial crisis. Everybody, both those who were involved prior to the financial crisis and the restructuring of CAT and those who have come to work at CAT since, acknowledge that CAT has radically changed, but evaluate these changes differently. For some, these changes have understandably been incredibly distressing at a personal level and are perceived as being detrimental to CAT as an organisation. Others perceive the changes as inevitable and necessary for CAT's survival but were very poorly implemented, while others are more noncommittal. My personal position is that organisations must adapt to both changing internal dynamics and external pressures, and failure to change in response to changing circumstances will lead to its demise. CAT had to face an extraordinary combination of circumstances, the most significant being a huge increase in the turnover and complexity of the organisation and a financial crisis that entailed a concomitant adoption of a more conventional style of management. While these changes have been distressing, at both a personal and organisational level, CAT has emerged from this traumatic nadir and continues to perform a valuable role in the environmental movement. Furthermore, the wider social, political and cultural context has also changed significantly since the 1970s–80s. In particular, environmental issues are no longer regarded as irrelevant or peripheral but are now centre stage. This has entailed that CAT and other environmental organisations have had to adapt and rethink the nature of their roles in this new context.

As I have indicated above, there is a longitudinal dimension to the research. This historical aspect is primarily based on an oral history project and supplemented by qualitative interviews that I have undertaken with past members of staff who are not represented in the oral history project. The oral history project was instigated by Allan Shepherd, who worked at CAT for 20 years. Shepherd and volunteers interviewed 100 people, mostly in 2012, creating an archive of 150 hours of individual reflections on CAT at various points in its history. A wide range of people were interviewed from early pioneers to more recent arrivals, from restaurant workers and builders to directors and trustees. There were also interviews with local people from the Dyfi Valley and some of the children who are often collectively referred to as the Quarry kids (see Shepherd 2015a and 2015b). Most of the recordings of the oral history interviews, along with a substantial archive of various documents, are now held at the National Library of Wales in Aberystwyth. These recollections constitute what Shepherd calls a fragmented 'collective memory of CAT', which give a glimpse of 'what it has been like to have been an environmentalist working on the margins of society these past forty years' (Shepherd 2015a: 36). My intent is to continue the work that Shepherd started. This is a vast untapped resource, and I hope that this monograph will inspire future researchers to use the archive to explore themes, such as inclusivity and sustainability, that I have not unpacked here.

While there is a historical aspect to the research, this book is not intended to simply be a history of CAT. Consequently, I have also interviewed members of the current senior management team, and members of the Board of Trustees, as well as current staff members. I have also attended several workshops and seminars organised as part of CAT's educational programme. To ensure clarity, I will use both first and last names when referring to the oral history project, as these interviews are available to the general public at the National Library. I list all the names and catalogue details of these publicly available recordings in the Appendix. On the other hand, when using informant testimony from the 20 interviews I conducted, I will simply use first names.

The third aspect of the methodology is to look at the rhetorical strategies used by CAT. All those who are convinced that we are facing an environmental crisis must persuade others. It is all very well to suggest achievable alternatives for addressing environmental concerns, but others must be convinced that these are preferable to the current ways of being. CAT has produced an abundance of communications – from flyers to complex reports – intended to persuade others that their practical solutions are not only achievable but also desirable. This rhetorical analysis will particularly examine what rhetoricians call *kairos*. This literally means time and 'indicates both the occasion for discourse and the surrounding conditions that present the rhetor with opportunities and constraints' (Longaker and Walker 2011: 9). Timeliness is perhaps the best way to understand the concept of *kairos* – what conditions prevail to make a particular argument persuasive or otherwise. At certain times, audiences will be more receptive than others. For example, CAT was founded in a particular *kairotic* moment that enabled its message to attract an audience. Arguments also appeal in different ways. Classical rhetoric suggests that there are three modes of appeal that constitute an argument. These modes of appeal derive from the three aspects of the communicative act: the speaker, the message and the audience. *Ethos* relates to the credibility of the speaker; *logos* indicates the strength of the argument itself; and *pathos* refers to the ability to appeal to the emotions of an audience.

While in classical rhetoric Aristotle suggests that the credibility of an argument 'should be achieved by what the speaker says, not by what people think of his character before he begins to speak' (Rhys-Roberts no date: 8), nonetheless prior impressions are significant when trying to convince a potential audience. Consequently, it is important to trace how CAT's credibility as an authoritative voice on environmental matters has changed over time. CAT, because it proposes practical solutions, tends to prioritise *logos* through its appeal to quantitative data and the use of the conventions of quasi-scientific idiom. The appeal to *pathos* tends to be rather attenuated in CAT's discourse. Nonetheless, it is clear that CAT's rhetoric does have an emotional appeal, which I argue is largely due to its history and the Quarry itself.

Continuity and Change

Both CAT as an organisation and the wider context in which it operates is very different from when CAT was founded. CAT has had to adapt to the challenges of the changing context. The most significant of these contextual changes over the five decades of CAT's existence have been:

1   There is now a much greater public awareness of environmental issues.
2   Many of the ideas, such as the use of renewable energy, that CAT advocated in the early decades have moved from the margins to the mainstream.
3   Environmental issues are now high on the agenda at global, national and local political levels.
4   Environmental issues are now considered newsworthy and are often represented in popular culture.
5   The main focus of environmental concerns has shifted from issues such as peak oil, use of pesticide and the depletion of ozone to the climate emergency and the loss of biodiversity.
6   The development of a green economy and employment market. In the first few decades of CAT's existence, there was almost no green sector, and CAT was one of the few employers that could offer employment opportunities for qualified engineers, builders, architects and so on, who were concerned with environmental issues. Now the green sector has increased exponentially, and many organisations, industries and government at national and local levels must consider sustainability.

In interviews with people who are involved with CAT today – workers, senior management and trustees – it is widely acknowledged that while CAT is a very different organisation from when it was founded, there is something that has remained consistent over the decades that somehow defines CAT's ethos. The Chair of the Trustees observes:

> Clearly CAT is a different organisation to what it was ten years ago and although it might have lost something along the way, what pleases me is that CAT is still visibly CAT.
>
> (Mick, interview May 2021)

Here he implies that CAT has a sort of inherent core that has remained unscathed despite the radical changes in the context and the internal dynamics of the organisation. He, like many others, remains unclear as to what this essential core of CAT might be. He also acknowledges that not everybody would perceive the changes in CAT in this way. I agree with Mick's assessment, and while there have been radical changes, there still is a consistent core that can be traced throughout CAT's history from the pioneers to the current day and is still the main focus of CAT's activities. This consistent core has three intertwined aspects:

1   To identify practical solutions to environmental problems.
2   To challenge the 'business as usual' model of addressing the environmental devastation caused by human activity.
3   To communicate and educate the public about the environmental devastation faced by humanity and the planet and informing as wide a public as possible about plausible solutions to address this destruction.

Discourses about environmental issues, what is meant by sustainability and how to address the crises have evolved significantly in the last decades. CAT has both been part of this shifting discourse and responded to it. Consequently, while, practical solutions, challenging the status quo and communication remain at the core of the CAT project, these central principles have been interpreted and implemented in different ways throughout CAT's history. In this case study, I identify a number of themes that can be traced throughout CAT's history, that are still current today and that can elucidate the challenges faced by those involved in trying to address the climate emergency, loss of biodiversity, pollution and other environmental issues.

Themes

The main themes that constitute the threads that relate CAT's experiences to more current debates about the environment include:

*   To what extent is sustainability a radical agenda? This has two intertwined aspects. Firstly, is environmentalism an oppositional ideology that demands systemic changes? Second, does sustainability necessarily entail major life-style changes?
*   Is the sustainability agenda best effected from within or outside the established political, social and economic structures?
*   Are the anthropogenic causes of environmental devastation best addressed at an individual or structural level?
*   What roles do science and technology play in both the destruction and the protection of the environment?
*   What are the best means of communicating, educating and inspiring individuals, governments and organisations to take action to address environmental crises?

Radical But Not Too Radical!

This theme considers the extent to which environmentalism challenges the status quo and predominant cultural values. Care for the environment might be informed by a radical ideological perspective such as deep ecology or simply be a concern about the local environmental impact of a proposed chicken farm in one's neighbourhood. Responses might include anything from being conscientious about recycling to protest and civil disobedience. Consequently, Andrew Dobson makes a distinction between environmentalism and ecologism. Dobson suggests that environmentalism is not an ideology as it proposes that environmental issues 'can be solved without fundamental changes in present values or patterns of production and consumption' (Dobson 2007: 2). On the other hand, ecologism is an ideology. Dobson proposes an ideology as having three basic features: an analytical description of society, a prescription for a better society and a programme for political action to transform society (Dobson 2007: 3). As an ideology, ecologism advocates 'radical changes in our relationship with the non-human natural world and in our mode of social and political life' (Dobson 2007: 3). For example, environmentalism might promote the use of electric cars, whereas ecologism would advocate radical changes in personal transport, such as demands to reduce if not abolish private cars and reducing the distances that we travel for work and leisure. If radicalism is understood as an agenda that demands major changes to social, political and economic structures, then ecologism would be considered radical whereas environmentalism would not be.

While most of the people involved in CAT lean towards the political left, they tend not to perceive their agenda as informed by a political ideology but are only focused on the imperative to find practical solutions to the environmental crises. Nonetheless, although often implicit rather than explicit, the logic of the pragmatic solutions advocated by CAT is ideological and tends towards ecologism rather than the more superficial approach of what Dobson terms environmentalism.

There is an inherent double bind in the radical characteristic of ecologism. On the one hand, ecologism proposes that the norms and practices fostered by the hegemony of neoliberal capitalism and the concomitant culture of consumerism are the root causes of environmental devastation. Therefore, ecologism directly challenges the dominant and taken for granted mores of everyday life, social patterns, economic structures and political processes. On the other hand, radical ideas can alienate those who are not yet convinced by ecologism's imperative to make comprehensive changes to lifestyles and society. Most of us are too comfortable with the familiar to accept any calls for radical changes to the status quo and personal lifestyle choices. There is a sense that increased consumption is indicative of success and that 'reducing one's consumption expenditures to live more sustainably translate into welfare losses so that following a green lifestyle would be viewed as involving sacrifice' (Binder and Blankenberg 2017: 305). In 2005, Murdo Fraser – a spokesman for the Scottish Conservative Party – stated: 'I don't think we should give house room to the mumbo-jumbo environmentalist clap-trap of the Green Party, who think we should go back to living in caves' (cited in Soper 2008: 583).

If one accepts the dominant idea that there is a direct relationship between qualifications, experience and remuneration, many of the people working for CAT did actually sacrifice a great deal, particularly in the very early pioneering days. Most of the staff were highly qualified and experienced in their various fields of expertise and would have been paid considerably more elsewhere. It is now widely accepted that the 'spend, spend' and 'more stuff' ideology of consumer capitalism, or what Fiona Brannigan (2015) calls 'the consumption-happiness myth' is not the only model of the good life. The majority of those who worked at CAT suggest that the relatively low pay and the concomitant lack of being able to acquire an abundance of material possessions were more than compensated by other aspects, such as working for the environmental cause itself, the beauty of the local environment, the camaraderie of the group and/or the control that they felt they had over their work.

The conundrum is how to get people to accept the radical changes that ecologism seems to entail. Two perspectives can help reconcile the need for radical changes and getting people to accept the need for change. First, the distinction between ecologism and environmentalism should not be understood as a binary opposition, but as a spectrum of views, sometimes articulated in terms of shades of green. Consequently, I use the term 'environmentalism' to indicate a discourse that signifies a wide range of views and practices – from deep ecologists to green consumers – that are informed by an understanding that we need to do more to protect the environment. This concern for the environment might manifest in a myriad of ways from taking a reusable shopping bag to the supermarket to civil disobedience. Second, radicalism must also be understood as a relational, shifting and gradated concept. What is considered an extremely radical change for some will be thought of as a relatively moderate adjustment for others. For example, cutting down on meat consumption might be radical for a committed carnivore, but for others, it might not necessarily be viewed as being a particularly extreme lifestyle choice. It is also clear that attitudes to vegetarianism, whether on ethical, health or environmental grounds, have shifted dramatically since the 1970s. In other words, what was considered radical yesterday might not be radical today. Indeed, some present and past members of staff do not consider CAT to be radical at all today, whereas others suggest that it still has a radical agenda.

The double bind of radicalism must be resolved by attempting an agenda that might be said to be 'radical, but not too radical' – to challenge without appearing to be inimical. Failure to challenge sufficiently will not generate a sufficient response to the environmental crises. Propose a too radical an agenda and many people will respond defensively and feel overwhelmed. This debate about the right balance between the changes needed to protect the environment and the imperative to appeal to others has been core throughout CAT's history. The balance of this challenge has shifted over the decades. In the early 1970s, the predominant ethos was that environmental issues were marginal at best. Consequently, CAT's agenda was often perceived as too radical and therefore outside many people's purview. Today, when environmental discourses have moved from the margins to the mainstream, there is perhaps an even greater challenge. Green ideas have become normalised and there is a danger that people get a sense that simply doing a bit of recycling or buying green products is adequate to address the environmental crises.

Inside or Outside?

This theme explores the debate about whether it is more effective to address environmental issues from within the socio-political structures or from the outside. The positioning of any environmental organisation in relationship to the wider society is linked to how radical its agenda is. This positioning is determined by where an oppositional organisation believes its agenda can be most effective. Broadly speaking, this positioning can be identified as inside or outside. The insider perspective perceives that change is best effected from within the established socio-political system. For example, the UK Green Party 'aims to create a just, equitable and sustainable society. We focus our efforts primarily, though not exclusively, through the electoral system' (Green Party 2022). On the other hand, other organisations argue that the systemic changes needed to protect the environment can only be effected from outside. Brian Doherty argues that the green movement, as a social movement, by definition must 'operate at least partly outside political institutions' (Doherty 2002: 10). For example, Extinction Rebellion advocates civil disobedience as the only effective strategy 'to transform the way in which we talk about the climate and ecological emergency and force governments all over the world to act' (Knights 2019: 9–10). Working outside established structures can be ineffective because of lack of social, political, cultural and economic power. However, working within the establishment can be ineffective because of pressures to make compromises that militate against making sufficient changes to address the environmental crises.

To some extent, environmentalism has come in from the cold as concern for the environment, in particular how to address the climate emergency, is now incorporated into mainstream discourses. This wider acknowledgement of environmental crises is ambivalent. On the one hand, it means that environmental movement organisations (EMOs) can be more effective in influencing legislation, policy and public opinion. Cary Coglianese observes that as a result of the wider recognition of environmental crises, 'the environmental movement has contributed to dramatic changes in law and in public values in the United States, and, as a result, society has achieved notable improvements in some of its underlying environmental conditions' (Coglianese 2001: 86). On the other hand, it could be argued that the dominant powers have simply appropriated green ideas without adopting any fundamental ideological changes. Consequently, it has become more problematic to challenge the underlying ideology of neo-liberal capitalism, which many perceive as the root cause of environmental degradation. Doherty (2002: 147) suggests that this has entailed that the environmental movement now 'concentrates on technical and scientific issues and no longer offers an alternative set of values or a challenge to the existing order'. Furthermore, with environmental issues becoming increasingly prevalent in the public domain and pressure for organisations to at least appear to be more sustainable, two interrelated phenomena have arisen, which militate against actually finding achievable alternatives to the environmental harms caused by human activity. These are what have come to be known as greenwashing and green consumerism. The journalist David Gelles observes:

> Greenwashing, when a company tries to portray itself as more environ-
> mentally minded than it actually is, has intensified in recent decades as
> consumers have warmed to sustainable and organic products and services.
> Brands trying to capitalize on that trend, often try to outdo one another
> with eco-credentials. However, in the rush to be seen as green, companies
> often exaggerate claims, or simply make things up.
>
> (Gelles 2015)

When asked about the name the Centre for Alternative Technology, many of my interviewees suggested that the concept of 'alternative' is not apposite as CAT no longer advocates alternatives as such, but that the name cannot be changed as it is a widely recognised and respected 'brand'. The term 'alterna-tive' seems to signify that mainstream politics, society and culture are failing to address the environmental crises, and that the solutions can only be found through resistant ideologies and practices. CAT has always had an ambivalent relationship with the establishment. CAT was founded by Gerard Morgan-Grenville who was in many ways a figure of the establishment, and his social standing and connections with the elite provided CAT a degree of support that it would otherwise not have garnered. For example, Morgan-Grenville got an endorsement from Roy Jenkins, a prominent member of Harold Wilson's Government, and other establishment figures for the nascent CAT. On the other hand, Morgan-Grenville also had an anti-establishment inclination. In some ways, CAT has reflected Morgan-Grenville's ambivalence towards the establishment. CAT's ability to exploit this ambiguity between insider and out-sider status accounts for its resilience. CAT in many ways has both establish-ment and countercultural credentials. While this has caused some tensions from time to time, it has enabled CAT to engage with a wide range of individuals and groups from royalty to Extinction Rebellion. CAT as a registered charity has always maintained its independence from Government and financed itself through a variety of means, such as charging an entrance fee to the visitor centre. This independence has been a significant aspect of CAT's success.

Individual Lifestyle or Systemic Changes?

This theme addresses the best level to effect changes – personal, local, national or global. While both individual and structural changes can be identified throughout CAT's history, the change of emphasis from the personal to the structural has tended to reflect the changing discourses in the wider environmental movement. Initially, the ethos of CAT was informed by the discourse of self-sufficiency, which focuses on personal and individual changes. In the UK, the concept of self-sufficiency was exemplified by John Seymour who ran a small farm in Pembrokeshire and published his influential book *Self-Sufficiency* in 1973. Many of the early members of CAT were influenced by Seymour's ideas, and indeed Seymour visited CAT. In the early decades, a smallholding was an important part of the visitor circuit.

Self-sufficiency is not simply about food. For example, in the early days of CAT, there were some debates about whether manufactured metal nails and so on should be used. Buying supplies from local builder's merchant can be said to not only support an exploitative capitalist economy, but the manufacturing and importing of materials from outside the region is also harmful to the environment. Consequently, rather than using metal nails, it would be much more environmentally sound to whittle wooden pegs. While pragmatism tended to prevail, nonetheless, reflection and debate about these issues are important and stimulated CAT to explore ways of building in more sustainable ways – for example, utilising locally sourced or reclaimed timber wherever possible in new builds. The debates between pragmatism and idealism have been a significant aspect of CAT's experience. Again, there are no clear-cut answers as to where the balance lies. Being too pragmatic fails to address the environmental crises in any substantive way but being too idealistic may appear to be impractical.

CAT was largely self-sufficient in power. All the electricity on-site was predominantly supplied by renewable sources. There were several small wind turbines that generated very little power. For example, one of the wind turbines that was on-site in the 1990s only generated 250 watts, not enough power for a modern kettle. In the 1980s, the whole site ran on about two and a half kilowatts. Even the Polenko turbine that was installed on a ridge above the Quarry in 1984 was largely symbolic. It was rated at 15 kilowatts, but because of its siting hardly ever produced that. CAT was never totally self-sufficient in power. Power was needed for cooking, and this was predominantly done with propane, which had to be bought in from outside. Furthermore, as the site is open to the paying public and CAT runs educational courses, in very dry spells when the hydro could not provide sufficient power and the batteries were depleted, there was an emergency diesel generator to ensure that the restaurant could continue serving and/or residential students were not plunged into darkness.

The technology of wind turbines has advanced well beyond the dreams of the early pioneers. When CAT started, there were no large-scale manufacturers of the type of wind turbines that people are familiar with today. Most of the electricity was provided by a small hydroelectric system. With the development of the hardware of AT and reappraising the situation, CAT no longer suggests that being self-sufficient in power is the most viable solution at an individual level. 'After decades of experimentation, research and thought, we don't think that disconnecting from the electricity grid is the green way forward' (Davis 2017). In the early 2000s, CAT itself realised that it would be beneficial to connect to the grid – partly because its energy needs could no longer be adequately met by the limited on-site supply; partly because it became possible to ensure that external suppliers were predominantly generating electricity from renewable sources; and partly because in times of relatively abundant power and low demand, CAT could sell energy back to the grid. In some ways, not being connected to the grid was counterproductive, in that it potentially could foster the idea that everyone should live off-grid. One of the most common criticisms levelled against CAT was that while it might be possible to live a sustainable lifestyle in a small community in a rural setting, much of what CAT advocates is not possible for individuals or nuclear families living in inner cities. While many people thought that there was much to admire at CAT, there was very little that they perceived could be readily adopted at home. Most people do not have space for wind turbines, let alone small hydro-schemes.

Self-sufficiency is also time consuming. Even if you have sufficient space for growing your own fruit and vegetables, you need to be preserving these in some way – making jams, pickling, canning and so on – to ensure food supply out of season, and most people lead very busy lives and have other demands on their time. Self-sufficiency also requires a certain practical skill set, which people do not necessarily have. It is doubtful that CAT would have progressed very far without the skills of someone like Bob Todd, who has a PhD in electrical engineering. The self-sufficiency agenda can in some senses be understood as essentially the preserve of a privileged few and not accessible for many.[5] Overall, CAT has moved away from the idea of self-sufficiency.

However, CAT has never been solely concerned with the idea of self-sufficiency. In 1977, Bob Todd in collaboration with other academics published *An Alternative Energy Strategy for the UK*. Todd recalls, this was a direct response to the criticism that:

> CAT was all terribly small scale and wasn't really going to make any impact on a country like Britain.
>
> (Bob Todd, Oral History)

Todd concluded that:

> You could over a period of maybe 25 years or 30 years, make a huge impact, both reducing the demand for electricity and supplying a large percentage of it from renewables.
>
> (Bob Todd, Oral History)

As a result, Todd was invited to present his ideas to various think tanks and groups within the government. This document formed the inspiration for CAT's more recent project Zero Carbon Britain, which focuses on large-scale systemic changes to achieve a carbon-free future.

Technology and Science

The term alternative technology (AT) signifies the ambivalence that the environmental movement has towards technology and science. The environmental movement tends to have a Manichean perspective on science and technology. There are the harmful and polluting technologies that are in large part responsible for the environmental crises – loss of biodiversity, climate emergency, depletion of finite resources and pollution. On the other hand, there is 'good technology', which is largely associated with renewable energy and is considered as much less damaging to the environment. Scale is often evoked to distinguish destructive as opposed to benign technology. Large-scale technology is associated with industrialisation, which is regarded as inherently exploitative of both humans and nature (see Dickson 1974: 41–62). On the other hand, to cite the title of E.F. Schumacher's (2011) seminal text, *Small is Beautiful*, AT, which is not based on the domination of humans and nature, is 'based on small-scale decentralized units' (Dickson 1974: 98). This evaluation of scale has to some extent been reassessed and is represented by the move from the 'Do It Yourself' (DIY) ethos of early AT, symbolised by the Cretan wood and cloth windmill which could barely generate enough power for a light bulb, to the development of a green economy, symbolised by the megawatt wind farms that can generate electricity for thousands of homes.

Science has two interconnected aspects in the discourses of environmentalism. First, science not only provides evidence for the destruction of the environment, but also constructs the issue. For example, science and scientific modelling is often a requirement to show environmental harms. George Myerson and Yvonne Rydin (1996: 90) observe that 'at the heart of the global warming issue is the recognition that the issue only exists because of the scientific measurements'. This is not to suggest that global warming is not real, but that it is only made visible through scientific measurement. Second, the idiom of science signifies authority. Rom Harré, Jens Brockmeier and Peter Mühlhäusler (1999: 51) describe scientific discourse as a 'potent rhetorical resource'. Lexical choices and grammatical constructions in what Carol Reeves calls 'the language of science' is 'an artifice that displays science as an objective enterprise' (Reeves 2005: 50). This artifice of objective fact is also supported by numerical data and by visual images, particularly graphs. For example, CAT's Zero Carbon Britain (ZCB) Reports are replete with graphs and technical vocabulary and frequently uses the passive voice. In other words, ZCB tends towards the conventions of scientific language to emphasise objectivity, rationality and plausibility. This use of a scientific idiom is integral to CAT's rhetorical strategy and its emphasis on practical solutions.

Communication and Education

It is of no use to establish a little eco-utopian community if all of humanity and the planet are still hurtling towards an environmental catastrophe. Consequently, communication and education have been core aspects of the environmental movement. It is clear that from the very beginning of CAT that dissemination of ideas was significant. The Articles of Association of CAT as a legally recognised charity state clearly that one of its prime objectives is:

> To amass or cause to be amassed information of relevance to environmental improvement and to *disseminate* such information.
>
> (Society for Environmental Improvement 1973)

This aspect has remained consistent throughout CAT's history. The most recent Articles of Association rewords this objective slightly and if anything enhances the communication and educational aspects.

> To advance the *education* of the public in subjects related to sustainable development and the protection, enhancement and rehabilitation of the environment, the use of natural resources and sustainable energy and to promote study and research in such subjects provided that the useful results of such study are *disseminated* to the public at large.
>
> (CAT 2021)[6]

CAT has tried to address as wide a constituency as possible, from the day visitor who perceives CAT as a tourist attraction to the pilgrims already well versed in the philosophy and practices of AT, and from primary school children on a school visit to students on one of the MSc courses on sustainability. This has always provided a challenge to CAT. Too much information alienates the unversed and too little information can antagonise the pilgrims.

This theme focuses on what has come to be known as environmental communication: the imperative to inform others about the environmental crises. However, what has been called the information deficit model of communication is no longer widely accepted. It is not sufficient to simply provide people with information about the environmental damage caused by human activity. Kari Marie Norgaard in her fascinating ethnographic study of a small Norwegian community observes that 'despite the fact people were clearly aware of global warming as a phenomenon, everyday life in Bygdaby went on as though it did not exist' (Norgaard 2011: xvi). Therefore, it is critical to persuade others, at whatever level, to act. Environmental communication is inherently rhetorical as it hopes to persuade others. Consequently, I draw on rhetorical analysis (Killingworth and Palmer 1992; Myerson and Rydin 1996; Ross 2017) and eco-linguistics (Stibbe 2015; Fill and Mühlhäusler 2001) to analyse the ways in which CAT disseminates its views on the environment and renewable energy.

CAT's Networks

While CAT is located in a fairly remote location, it has never been isolated. CAT can be considered as a central node in a variety of networks. There is what might be considered a meta-network, a network of networks so to speak, of all environmental groups and organisations. This meta-network has clearly expanded over the decades since CAT was established. For example, Extinction Rebellion founded in 2018 might be considered as both a new node in this meta-network of environmental organisations and also a network of its own. While a number of people either came from or went on to work for organisations such as FoE, CAT's formal links with this meta-network in terms of sharing and collaboration are minimal. Nonetheless, CAT is well known and has a good reputation in environmental networks nationally (both Welsh and British) and internationally. In recent years, it has linked into the environmental network in a more substantive way. For example, CAT has recently joined with five other organisations (Extinction Rebellion, the Woodland Trust, Whale and Dolphin Conservation, CAFOD and the Future Generations Commission for Wales) to demand that all political parties prioritise tackling the climate emergency in their manifestos (CAT 2021).

I will not be looking at this metalevel in any detail. However, CAT is linked into three other networks. I will term these networks – the direct, the local and the diffuse. The direct network includes a small number of immediate spin-offs. These include the Quarry Shop – a café and wholefood store that was set up on the high street in Machynlleth; Dulas Ltd – a renewable energy company; Undergrowth Housing Cooperative that bought a nearby property to provide low-cost and low-environmental impact accommodation and The Urban Centre for Appropriate Technology (UCAT) in Bristol.[7] The local network includes local green initiatives that CAT has been involved in, as well as the wider social, cultural and economic impact CAT has had on the Dyfi Valley. This network includes Bro Dyfi – a community renewable energy project; Ecodyfi – a group that supports and fosters a greener community; and the Dyfi Biosphere – a UNESCO designation that promotes sustainability in the Dyfi valley. This local network is at least partly due to CAT acting as a sort of lodestone, attracting people to the area who are interested in green issues. For example, there are quite a few people who came to CAT as volunteers or who left CAT's employment and have remained in the Dyfi valley region. The diffuse network is perhaps the most significant. Over the decades, numerous people have been inspired by CAT. These individuals have taken ideas, practices and skills that they have garnered during their time at CAT, whether for a few hours or over many years, to other parts of the world. The most obvious manifestation of this diffuse network is the CAT Membership scheme. This diffuse network has been particularly strengthened by the development of the Graduate School of the Environment (GSE), which offers MSc courses in various aspects of sustainability. Many graduates of the GSE go on to work in the green economy, start their own business or act as consultants for other companies. These networks indicate that CAT is a much more significant player in promoting the protection of the environment than is perhaps immediately apparent.

Structure of the Book

Chapter 1: The Context and Early History of the Centre for Alternative Technology

This chapter briefly outlines the historical context of the 1970s, and the emergence of the modern environmental movement. The development of modern forms of environmentalism is a response to four major factors: a growing ambivalence about science, technology and industrialisation; an awareness of the Earth as a single place; the counterculture; and the publication of popular texts on environmental issues. In this chapter, I will explore the idea of AT and how it captures the idea of practical utopianism and conveys a sense of being radical, but not too radical. The chapter explores two main points in the history of CAT: the pioneering days and the changes that CAT made in the 1990s referred to as Gearchange. The early pioneering days were informed by the idea of self-sufficiency and E.F. Schumacher's philosophy that 'small is beautiful'. However, with the development of the idea of sustainable development, the improvement of renewable technology and many of CAT's ideas becoming more widely known, the small is beautiful ethos was no longer apposite in the context of what Stephen C. Young (2000) has called 'ecological modernisation'. In the early 1990s, CAT saw the opportunity to reach out to a wider public. This period culminated with the building of a water-powered cliff railway, which would make access to the site easier, demonstrate AT and be a visitor attraction in its own right. Finally, in the early 2000s, CAT connected to the grid, which symbolised the demise of the self-sufficiency model and the dominance of the idea of sustainability.

Chapter 2: From Community to Network

This chapter explores the community aspect of CAT. In the early days, the community was a significant aspect of CAT. However, there was an ongoing debate as to the purpose of the community. Some thought that it was simply a convenient place to live, others thought that the community was essential to the project as an experiment of living with renewables and being as self-sufficient as possible. Yet others thought that the community was integral to CAT's agenda to persuade others of the viability of a sustainable lifestyle. However, the success of the rhetorical aspect of the community was equivocal. On the one hand, it was a relatively small community in a rural context, and the lifestyle therefore was not relevant to mainstream society. On the other hand, demonstrating that it was possible to live a sustainable lifestyle added to the credibility of CAT's argument. This chapter also examines in what sense CAT was a community *per se* as not everybody who was involved in CAT lived on-site. Even in the 1980s, when the site community was at its strongest, only half the people who worked at CAT actually lived on-site. Nonetheless, CAT produced strong social bonding capital amongst the whole staff body through a sense of sharing. However, as time progressed, CAT's focus shifted from creating strong social bonding capital to emphasising bridging social capital. With the emergence of ecological modernisation, the growing size and ambition of CAT, advances in the technology of renewable energy and the concomitant anachronism of the small is beautiful ethos, the significance of the site community became increasingly attenuated until it folded in around 2010. Nonetheless, while CAT now resembles a more conventional non-governmental organisation (NGO), strong social bonding capital, a sense of community and sharing were absolutely integral in the formation of CAT's networks of influence.

Chapter 3: Decision Making

This chapter explores the changes in the decision-making and managerial structures of CAT. In many ways, this chapter extends the argument of the previous chapter as the changes in the decision-making processes mirror the significance of the community over time. In the very early days, there was some tension between the fairly autocratic authority of CAT's founder Gerard Morgan-Grenville and the more *ad hoc* decision making of pioneers living on-site. When Morgan-Grenville appointed Roderick James in 1975 as CAT's first Director, a balance was struck between having some formal structure to the decision making and enabling everybody to feel like they had an input into the decision making. The overall ethos was to foster a sense of egalitarianism through consensus. This ethos fostered strong social bonding capital. It can also be understood as a form of prefigurative politics that reflected the idea that an environmentally sustainable society would also be socially egalitarian. This egalitarian aspiration tended to prioritise the purity of means over the primacy of ends. However, consensus decision making and egalitarianism were not sustainable in the long run and collapsed due to both internal and external pressures. Internally, it seemed that as CAT grew in size and complexity as an organisation, consensus decision making became increasingly unwieldy. Externally, with the growing awareness of the urgency of addressing the climate change, primacy of ends had to take priority over the purity of means. The final nail in the egalitarian experiment was that CAT also experienced a very severe financial disaster, which meant that a more conventional managerial structure had to be instituted in 2010. Although this was a very traumatic experience in CAT's history and those who were strong advocates of the participatory and egalitarian ethos felt badly betrayed, this change was consistent with the need to focus more on bridging social capital.

Chapter 4: Communication, Education and Persuasion

This chapter investigates the challenges that CAT faced in trying to convince others about the environmental devastation caused by human activity and the practical solutions that are required to redress that destructive tendency. There is, after all, no point in a small group in rural Wales living sustainably if the rest of the humanity continues to pollute the planet and exhaust all its limited resources. Consequently, there is an imperative to persuade others of the need to adopt a more environmentally friendly lifestyle. This chapter draws on rhetorical analysis to assess CAT's communication and educational strategies. The problem faced by all environmental advocates is that they must persuade a very diverse audience from policy makers to school children; from those who are indifferent to those who are hostile to the message. In 1975, CAT opened to the paying public as a visitor centre. In many ways, CAT's location in an area that attracted tourists was fortuitous. Tourists in the area often included CAT as one of their holiday activities. CAT hopes that the visitor circuit will inspire people to look more deeply into the environmental issues and plausible solutions. To this end, CAT offers a range of courses on various aspects of sustainability. This culminated with the founding of the GSE that offers MSc courses on sustainability. While you are never going to persuade everyone, there is a good deal of evidence that many people have found CAT an inspiring place, whether they came as a child whilst on holiday with their parents or took a graduate course. In this way, CAT's influence and impact has rippled out, albeit in a relatively invisible way, from its remote rural location.

Chapter 5: Zero Carbon Britain

This chapter investigates CAT's response to the criticism that the practical solutions it suggested were too small and therefore were not apposite on a larger scale. CAT's first serious attempt to address this criticism was in 1977 with the publication *An Alternative Energy Strategy*. This publication extrapolated from CAT's experience and from quantitative data to suggest that the UK as a nation could generate the majority of its energy requirements through renewable resources. In the early 2000s, it became apparent that climate change overshadowed all other environmental crises. It was also apparent that *An Alternative Energy Strategy* was out of date. Consequently, CAT began a new project that took the basic ideas of *An Alternative Energy Strategy*, CAT's experience and data from outside sources to see if it was plausible for the UK to reduce its carbon emissions to zero. In 2007, CAT published the first *Zero Carbon Britain Report*. CAT was therefore well ahead of the game in proposing that zero carbon was not only necessary but also plausible. The focus of this chapter is an investigation of the rhetorical style of *An Alternative Energy Strategy* and the first *Zero Carbon Britain Report*. Since this first report, CAT has published a further three reports and two supplementary reports. Perhaps most importantly for CAT was the establishment of an Innovation Lab in 2020, which aims to facilitate councils and organisations to decarbonise their activities.

Conclusion

In the conclusion, I argue that although CAT might initially seem to be an obscure and maverick organisation located in a relatively remote part of the UK, its impact has been much more significant than first appears. CAT is not only a credible voice in the environmental movement, but also its focus on identifying practical solutions to environmental crises is more important than ever. My overall conclusion is that we need as many practical utopian discourses as possible, not only to identify plausible solutions to the climate emergency, but also to overcome the paralysis of an eco-anxiety caused by an overwhelming sense that humanity and the planet faces an eco-apocalypse.

Notes

1 Roderick James was CAT's first director. This comment was recalled in an oral history interview. See the Methodology section for details about the oral history project.
2 While a few in the local Welsh community in the 1970s did perhaps perceive CAT as 'a barbaric intrusion', this was largely because most of the CAT pioneers were English and many were not conscious that there is a distinctive Welsh culture.
3 Just over half of the TripAdvisor reviews were positive, and most of these were very positive. This is probably not a fair sample as Wales has just emerged from the coronavirus disease–2019 (COVID-19) lockdown, and the site has largely been closed or had very restricted access over the last 18 months.
4 It was eventually sold to a private company in 2011 (see BBC News www.bbc.co.uk/news/uk-england-south-yorkshire-12824262). It is now an outdoor activity centre, which uses some of the renewable energy resources, such as a bank of solar panels. The company promotes what is now called The Dearne Valley Activity Centre as 'our most environmentally sustainable centre' (www.kingswood.co.uk/activity-centres/dearne-valley).
5 It is noticeable that in the UK the environmental movement has largely been dominated by the white middle classes, and only comparatively recently has started to consider sustainability in relation to inclusivity and diversity.
6 My emphasis. Note also that the charity was first incorporated by Gerard Morgan-Grenville under the name The Society for Environmental Improvement. This name was legally changed in 1990 to The Centre for Alternative Technology Charity Ltd.
7 Now called the Centre for Sustainable Energy www.cse.org.uk

References

Binder, M. & Blankenberg, A. (2017). Green Lifestyles and Subjective Well-Being: More About Self-Image Than Actual Behaviour? *Journal of Economic Behaviour and Organization* 137, 304–23. doi: 10.1016/j.jebo.2017.03.09

Brannigan, F. (2015). Dismantling the Consumption-Happiness Myth: A Neuropsychological Perspective on the Mechanisms That Lock Us in to Unsustainable Consumption. In L. Whitmarsh, S. O'Neill & I. Lorenzoni (Eds.), *Engaging the Public With Climate Change: Behaviour Change and Communication*. London: Routledge, 84–99.

Centre for Alternative Technology (CAT). (2021). Memorandum of Association. Available at: https://find-and-update.company-information.service.gov.uk/company/01090006 (Accessed 7 June 2021).

Coglianese, C. (2001). Social Movements, Law and Society: The Institutionalization of the Environmental Movement. *University of Pennsylvania Law Review* 150(1), 85–118. doi: 10.2307/3312913

Crompton, C.T. (1967). The Treatment of Waste Slate Heaps. *Town Planning Review* 37(4), 291–304.

Davis, M. (2017). On-Grid or Off-Grid? Which is Greener. Centre for Alternative Technology. Available at: https://cat.org.uk/grid-off-grid-greener (Accessed 3 January 2022).

Dickson, D. (1974). *Alternative Technology and the Politics of Technical Change*. Glasgow: Fontana Collins.

Dobson, A. (2007). *Green Political Thought*. London: Routledge (4th edition).

Doherty, B. (2002). *Ideas and Actions in the Green Movement*. London: Routledge.

Fill, A. & Mühlhäusler, P. (Eds.) (2001). *The Ecolinguistics Reader*. London: Continuum.

Gardner, Z. (2008). *Landscapes of Power: The Cultural and Historical Geographies of Renewable Energy in Britain Since the 1870s*. PhD Thesis: University of Nottingham.

Gelles, D. (2015). Social Responsibility That Rubs Right Off. *The New York Times*. (17 October) Available at: www.nytimes.com/2015/10/18/business/energy-environment/social-responsibility-that-rubs-right-off.html (Accessed 3 January 2022).

Green Party (2022) *Green Party Policy*. Available at: https://policy.greenparty.org.uk (Accessed 23 June 2022).

Gregg, S.E. & Chryssides, G.D. (2019). Relational Religious Lives: Beyond Insider/Outsider Binaries in the Study of Religion. In S.E. Gregg & G.D. Chryssides (Eds.), *The Insider/Outsider Debate: New Perspectives in the Study of Religion*. Sheffield: Equinox, 3–29.

Haraway, D. (2004). Situated Knowledge: The Science Question in Feminism and the Privilege of Partial Perspective. In S. Harding (Ed.), *The Feminist Standpoint Theory Reader: Intellectual and Political Controversies*. London: Routledge, 81–102.

Harré, R., Brockmeier, J. & Mühlhäusler, P. (1999). *Greenspeak: A Study of Environmental Discourse*. London: Sage Publications.

Intergovernmental Panel on Climate Change (IPCC). (2021). *6th Assessment Report*. Available at: www.ipcc.ch/assessment-report/ar6 (Accessed 16 August 2021).

Killingworth, M.J. & Palmer, J.S. (1992). *Ecospeak: Rhetoric and Environmental Politics in America*. Carbondale: Southern Illinois University Press.

Kirk, A.G. (2007). *Counterculture Green: The Whole Earth Catalog and American Environmentalism*. Lawrence: University Press of Kansas.

Knights, S. (2019). Introduction: The Story So Far. In Extinction Rebellion (Ed.) *This is Not a Drill: An Extinction Rebellion Handbook*. London: Penguin, 9–13.

Levitas, R. (2011). *The Concept of Utopia*. Oxford: Peter Lang.

Levitas, R. (2013). *Utopia as Method*. Basingstoke: Palgrave Macmillan.

Longaker, M.G. & Walker, J. (2011). *Rhetorical Analysis: A Brief Guide for Writers*. London: Longman.

Mathews, F. (1995). Conservation and Self-Realization: A Deep Ecology Perspective. In A. Drengson & Y. Inoue (Eds.), *The Deep Ecology Movement: An Introductory Anthology*. Berkley: North Atlantic Books, 124–135.

Mojo. (2021). Inspiring and Beautiful Family Day Out. *TripAdvisor*. Available at: www.tripadvisor.co.uk/Attraction_Review-g552064-d295844-Reviews-or30-Centre_for_Alternative_Technology-Machynlleth_Powys_Wales.html (Accessed 23 June 2022).

Muffinn2. (2021). A Massive Opportunity to Educate and Inspire Missed at Almost Every Level. *TripAdvisor*. Available at: www.tripadvisor.co.uk/Attraction_Review-g552064-d295844-Reviews-or30-Centre_for_Alternative_Technology-Machynlleth_Powys_Wales.html (Accessed 23 June 2022).

Myerson, G. & Rydin, Y. (1996). *The Language of Environment: A New Rhetoric.* London: UCL Press.

Naess, A. (1995). The Shallow and the Deep, Long-Range Ecology Movement: A Summary. In A. Drengson & Y. Inoue (Eds.), *The Deep Ecology Movement: An Introductory Anthology.* Berkley: North Atlantic Books, 3–10.

Niesewand, N. (1998). Architecture: Will the Earth Move Anyone? *The Independent.* (11 December) Available at: www.independent.co.uk/arts-entertainment/architecture-will-the-earth-move-anyone-1190607.html (Accessed June 23 2022).

Norgaard, K.M. (2011). *Living in Denial: Climate Change, Emotions and Everyday Life.* Cambridge, MA: MIT Press.

Oltermann, P. (2021). Angela Merkel Says That Germany Must Do More to Fight Climate Crisis. *The Guardian* (18 July). Available at: www.theguardian.com/world/2021/jul/18/angela-merkel-to-visit-flood-ravaged-areas-in-germany (Accessed 11 August 2021).

O'Neill, S. & Nicholson-Cole, S. (2009). "Fear Won't Do It": Promoting Positive Engagement With Climate Change Through Visual and Iconic Representations. *Science Communication* 30(3), 355–79. doi: 10.1177/1075547008329201

Pepper, D. (1991). *Communes and the Green Vision: Counterculture, Lifestyle and the New Age.* London: Green Print.

Reeves, C. (2005). *The Language of Science.* London: Routledge.

Rhys-Roberts, W. (Trans.) (no date). *Aristotle: Rhetoric* Available at: www.bocc.ubi.pt/pag/Aristotle-rhetoric.pdf (Accessed 13 January 2022).

Ross, D.G. (2017). Introduction. In D.G. Ross (Ed.), *Topic Driven Environmental Rhetoric.* London: Routledge, 1–21.

Roszak, T. (1995). *The Making of a Counter Culture.* Berkley: University of California Press.

Sargisson, L. (2000). *Utopian Bodies and the Politics of Transgression.* London: Routledge.

Sargisson, L. (2012). *Fool's Gold: Utopianism in the Twenty-First Century.* Basingstoke: Palgrave Macmillan.

Schumacher, E.F. (2011). *Small is Beautiful: A Study of Economics as if People Mattered.* London: Vintage Books.

Shepherd, A. (2015a). Voices from a Disused Quarry: Change Making, Class Dynamics and Technological Experimentation at the Centre for Alternative Technology. *Oral History* 43(2), 35–49.

Shepherd, A. (2015b). *Voices from a Disused Quarry: An Oral History of the Centre for Alternative Technology.* Machynlleth: CAT Publications.

Smales, J. (1999). The Right of Reply. *The Independent* (12 December). Available at: www-proquest-com.ezproxy.wlv.ac.uk/docview/311624101?OpenUrlRefId=info:xri/sid:primo&accountid=14685 (Accessed 9 August 2021).

Smith, A. (2004). An R&D Lab for Utopia? Alternative Technology Centres in the UK. *The European Consortium for Political Research.* Available at: https://ecpr.eu/Events/Event/PaperDetails/13947 (Accessed 30 December 2021).

Society for Environmental Improvement Ltd. (1973). Memorandum and Articles of Association. Available at: https://find-and-update.company-information.service.gov.uk/company/01090006/filing-history (Accessed 7 June 2021).

Soper, K. (2008) Alternative Hedonism, Cultural Theory and the Role of Aesthetic Revisioning. *Cultural Studies* 22(5), 567–87. doi: 10.1080/09502380802245936

Stibbe, A. (2015). *Ecolinguistics: Language, Ecology and the Stories We Live By.* London: Routledge.

Urry, J. (2011). *Climate Change and Society.* Cambridge: Polity Press.

Wright, E.O. (2010). *Envisioning Real Utopias.* London: Verso.

Young, S.C. (Ed.) (2000). *The Emergence of Ecological Modernisation.* London: Routledge.

Zelko, F. (2013). *Make It A Greenpeace: The Rise of Countercultural Environmentalism.* Oxford: Oxford University Press.

1 The Context and Early History of the Centre for Alternative Technology

Introduction: The Historical Context

The Centre for Alternative Technology (CAT) emerged in a particular historical context, which Joachim Radkau (2014) has called 'the age of ecology', which he suggests began around 1970. He indicates that:

> Most elements of today's 'environmentalism' have had a long history under various other names ... One thing can be stated at once though: the networking, wide impact and global horizons that developed from 1970 on were more or less new.
>
> (Radkau 2014: 7)

While considerable ink has been spilt in debates about whether this was indeed something new, what distinguishes modern environmentalism from its previous iterations and when to date the origins of this new phenomenon, I agree with Radkau that the environmentalism that emerged in the late 1960s and early 1970s onwards is distinct from earlier concerns for the environment.

This paradigm shift can be described as the transformation from conservation to environmentalism. Conservation is perhaps best epitomised by John Muir (1838–1914) who was a mountaineer, writer and activist.[1] Muir was a founder member of the Sierra Club, which was founded in 1892 to protect the Sierra Nevada Mountains as a wilderness area. Joachim Radkau characterises Muir as 'the American prophet of national parks and the wilderness' (Radkau 2014: 36). While there is an implicit of critique of some aspects of modernity in Muir's writing, his focus was on the preservation of wilderness spaces. Muir perceived the wilderness as a therapeutic space and antidote to the stresses of modern life. In 1901, in a treatise *Our National Parks*, Muir wrote:

> Thousands of tired, nerve-shaken, over-civilized people are beginning to find out that going to the mountains is going home; that wildness is a necessity; and that mountain parks and reservations are useful not only as fountains of timber and irrigated rivers, but as fountains of life.
>
> (Muir 2019: 430)

DOI: 10.4324/9781003207702-2

On the other hand, the new environmentalism, of which alternative technology (AT) as well as groups such as Friends of the Earth (FoE) and Greenpeace are exemplars, tended to make a much more explicit link between environmental issues and the structures and ideology of modern capitalist and industrialised society. Brian Doherty (2002: 29) observes that this new form of environmentalism is new precisely because it 'linked environmental problems to structural features of western society'. For example, David Dickson describes AT as a technology that:

> Would embrace the tools, machines and techniques necessary to reflect and maintain non-oppressive and non-manipulative modes of social production, and a non-exploitative relationship with the natural environment.
> (Dickson 1974: 11)

There were four interconnecting contextual factors that created the conditions for the emergence of modern environmentalism: ambivalence about science, technology and industrialisation; the perception of planet earth as a single place; the counterculture; and popular publications on environmental sciences. None of these were new *per se*, and all have identifiable antecedents, but together they created the conditions in the 1960s and 1970s for the emergence of modern environmentalism, of which CAT is a prime example.

Ambivalence About Science, Technology and Industrialisation

There was an increasing ambivalence about science, technology and industrialisation post World War II. Undoubtedly science, technology and industry have brought an increase in wealth, convenience and comfort to many. However, there was also a growing awareness of the dark side of science and technology raised by events such as the thalidomide scandal of the late 1950s and early 1960s, when a commonly prescribed drug for anxiety and morning sickness caused major birth defects. There was anxiety not only about the destructive power of nuclear weapons, but also about the dangers of nuclear power. There was also growing recognition of the cost of industrialisation, both in terms of pollution and an understanding that the resources to power industrialisation and modern consumer lifestyles are not limitless. Barbara Ward, in her important book *Spaceship Earth*, published in 1966, wrote that the world is 'being driven onward at apocalyptic speed by science and technology' (Ward 1966: 1).

Anxiety about the pollution caused by industrial processes and modern lifestyles was not new. For example, air quality was a particular concern in Victorian Britain. Poor air quality was caused by burning coal, both domestically and for industry, as well as various chemical processes such as the manufacture of alkali. This led to the passing of various statutes, such as the Smoke Prohibition Act of 1821 (see Clapp 1994: 13–38). However, there were several events in the late 1960s and early 1970s that came to represent the environmental risks of industrialisation. This disquiet about the cost of industrialisation is best epitomised by the fire on the Cuyahoga River in Cleveland in 1969, which was caused by a spark igniting the polluted river. It has become an iconic event, at least in part because *Time* magazine published an article with a dramatic image of a tugboat engulfed in flame. It has been suggested that this event was the catalyst for the environmental movement and the stimulus for Richard Nixon creating the Environmental Protection Agency (EPA) in 1970. It is of course a much more complex narrative, not least because the image used by *Time* was of a fire from 1952 on the Cuyahoga River. Nonetheless, the fire in 1969 became an iconic event that signified the worst aspects of industrial pollution. In the United Kingdom (UK), an equivalent iconic event was when the tanker the Torrey Canyon hit the rocks off the Cornish coast in 1967 and more than 10,000 tons of crude oil spilled and washed up on the beaches of Cornwall, Guernsey and Brittany. The media coverage was replete with images of dying seabirds thickly coated with black and a huge army of volunteers wading through the sludge on what had previously been pristine beaches. Phil Macnaghtan and John Urry suggest that the Torrey Canyon oil spill was 'a symbolic landmark event in the emergence of the UK environmental discourse' (Macnaghtan and Urry 1998: 49).

Concerns about the depletion of resources were not new either. For example, at the end of the 18th century, there was a concern across Europe about the depletion of forests. Joachim Radkau identifies that around 1800 the term 'sustainability' (*Nachhaltigkeit*) became a magic word in German forestry', which referred to the realisation that felling timber had to be limited to the forest's ability to regenerate (Radkau 2014: 17). This realisation informed the idea of forest management.

In the early 1970s, concern about resources, particularly oil, tended to be geo-political and economic, rather than environmental. In 1973, the major oil producing nations under the banner of the Organization of the Petroleum Exporting Countries (OPEC), which except for Venezuela were all in the Middle East, initiated an oil embargo on countries including the USA and the UK, which had supported Israel in the Yom Kippur war. This embargo led to a massive rise in oil prices and petrol rationing in the USA and the UK. While this oil shortage did not directly stimulate any major investment in renewable energy, it did contribute to a realisation that oil could not be squandered without thought. Sami Alpanda and Adrian Peralta-Alva (2010: 825) argue that 'the oil crisis spurred the adoption of energy-saving technologies'. The National Center for Appropriate Technology, located in Butte, Montana, was explicitly 'created during the oil crisis of the 1970s to develop inexpensive energy-saving strategies for low-income communities' (NCAT 2021).

This energy crisis coincided with the narrative of peak oil. This idea, first posited by M.K. Hubbert in the 1950s, suggests that production of oil, which is a non-renewable resource, is shaped like a bell curve, which reaches a peak in production, and when approximately half the reserves have been extracted, the production will gradually decline to zero, and that there would be a concomitant rise in price. While this is a much debated and disputed concept, it does seem to make logical sense. The theory of peak oil led to a consideration of what to do when either oil reserves were exhausted or became so problematic to extract that it becomes economically unviable (see Bardi 2009). Although, to a certain extent, concerns about the depletion of oil were at least temporarily undermined by the discovery and exploitation of North Sea oil.

Only One Planet Earth!

The second contextual factor for the development of modern environmentalism was the increasing consciousness, kindled by images of earth from space, that we live on a single planet. This awareness reinforced concerns about limited resources and the cost of industrial pollution. The first widely circulated photograph of planet earth from space was taken by William Anders in 1968 while orbiting the moon in Apollo 8. This image shows part of the Earth rising above the lunar landscape, with a clear divide between the nocturnal and daytime sides of the planet. Neil Maher suggests that the inclusion of the lunar landscape in this image visually signified John F. Kennedy's rhetoric of the extension of the frontier trope to space. However, in 1972, NASA released another photograph, this time taken from Apollo 17, that quickly replaced the Earthrise photograph. This image, because the sun was behind the spacecraft, shows planet earth illuminated as a perfect sphere, partially obscured by white clouds. Deep blue oceans, the reddish-brown land mass of the horn of Africa and the Middle East and the pristine white Antarctic are clearly identifiable. This image, sometimes referred to as the Blue Marble, is very aesthetically pleasing, and was published in papers around the world; Nicholas Mirzoeff (2015: 3) observes it is 'believed to be the most reproduced photograph ever'. Maher (2004: 529) suggests that the Blue Marble image represents not the new frontier of the Earthrise image, but 'an environmentally threatened home'. The Blue Marble image visually emphasises the idea of the planet as a single finite place. It reinforced the notion that resources, such as fossil fuels, are not inexhaustible and that we must live within our limits without poisoning this clearly finite home. This idea of the world as a single bounded place was reinforced by a plethora of titles and metaphors, such as spaceship earth, a global village and lifeboat earth.

   This idea of earth as a single place is extended by James Lovelock's Gaia hypothesis, named after the Greek Goddess of the Earth. This theory suggested that organic life and the environment interact as a self-regulating system analogous to a single organism. Lovelock had been working on this idea since 1960s and had published two articles in academic journals, first in 1972 and then in 1974. However, the Gaia hypothesis spread beyond a specialised academic community when he published *Gaia: A New Look at Life on Earth* in 1979. Lovelock deliberately wrote this book in a style that was accessible to a wider non-specialist audience, and it quickly became a best seller. While controversial, Lovelock's theory reinforced the idea that action in a specific location could have a global impact, as it could potentially disturb the balance of the planetary system. The Gaia hypothesis substantiated the call to 'think globally, act locally', which was the slogan that launched FoE in 1969, and has since become a mantra of the environmental movement more generally. The Gaia theory, while still controversial, has also informed debates about climate change. Perhaps the most significant theme in Lovelock's thesis was the idea of feedback loops, which is a significant concept in discourses about climate change. Eileen Crist and H. Bruce Rinker argue:

> Gaia theory proposes that organisms inflicting damage to their environment will eventually reap harsh consequences when feedback comes back to haunt them. We are currently experiencing such feedback in the form of climate change.
>
> (cited in Shamsuddha 2017: 68)

Human activity, such as creating greenhouse gas (GHG) emissions, leads to global warming, which itself produces more GHG. In other words, there is a point at which the GHG produced by human activity disturbs the self-regulating balance of the global system, known as homeostasis, to such an extent that you get positive feedback loop. This idea also informs another major metaphor in the environmental discourses – namely, that of a 'tipping point'.

The Counterculture and the Environment

The counterculture can be understood in a general and specific sense. In the general sense, it can refer to any cultural movement that challenges the dominant political, social and cultural values of the day. However, the counterculture is more frequently perceived as a very specific movement of the 1960s, and often referred to as the hippies. As I have already argued, although CAT was sometimes represented as being hippie, it never was part of the 1960s counterculture. Even the well-respected journalist Roger Harrabin, who is the BBC's correspondent on the environment, wrote in an article in 2014 for the *Guardian*: 'Welsh hippies ushered in an era of sustainable living well before the world had woken up to climate change' (Harrabin 2014). This is just sloppy journalism. The majority of people at CAT were not Welsh, certainly cannot be identified as hippies, and the hippie movement really had more or less vanished by the early 1970s. Consequently, any association of the environmental movement in general and CAT in particular with the counterculture is unhelpful at best.

Nonetheless, in the very early days, there was some imbrication between the counterculture and modern environmentalism, at least in appearance. Frank Zelko comments that 'the new Greenpeace-inspired environmentalists wore tie dyed tee shirts and long hair' which was in stark contrast to the conservationist style of the Sierra Club hikers of 'corduroys and cardigans' (Zelko 2013: 5). There were also some individuals and groups within the 1960s counterculture who were concerned with environmental issues. For example, the San Francisco Diggers, an important and politicised group that was part of the

California countercultural scene, wrote in their 1968 publication *The Digger Papers*:

> Industrialization was a battle with 19th century ecology to win breakfast at the cost of smog and insanity. Wars against ecology are suicidal. The U.S. standard of living is a bourgeoise baby blanket for executives who scream in their sleep. No Pleistocene swamp could match the pestilential horrors of modern urban sewage.
>
> (The Diggers 1968)

The environmental concerns of the counterculture were just a small part of a wider disillusionment with urban, consumer and industrialised life of main-stream society. This disillusionment fed into the back-to-the-land movement in the USA (see Jacob 1997) and to a certain extent in the UK (see Halfacree 2006). While in the USA, back-to-the-land tended to be framed in terms of homesteading and the idea of the frontier, in the UK, back-to-the-land was articulated in terms of smallholding and self-sufficiency, which was largely inspired by John Seymour's publications *Self-Sufficiency* (1973) and the beauti-fully illustrated and expanded *The Complete Book of Self-Sufficiency* (1976). The idea of self-sufficiency was also popularised in the BBC sitcom *The Good Life*, which ran from 1975 to 1978. In this comedy, the central characters, Tom and Barbara, attempt to become self-sufficient in their suburban house. While CAT cannot be considered as part of the back-to-the-land movement as it was a much more ambitious project, self-sufficiency was an important ideal in the vision and the early pioneering days. Gerard Morgan-Grenville in his vision statement for CAT wrote that there is 'a growing need for a more self-sufficient lifestyle' (Morgan-Grenville 1973). However, it was quickly realised at CAT that self-sufficiency was not the panacea for the environmental threats to the planet.

Perhaps, the most significant countercultural figure for the Alternative Technology Movement was Stewart Brand, who was the instigator of and driving force behind the *Whole Earth Catalog*. The first edition was published in 1968 and then quarterly until 1972. It sold millions of copies and won the National Book Award in 1972. It was then published intermittently until the last edition in 1994. It was no coincidence that the covers of all the *Whole Earth Catalogs* had a picture of earth from space. Stewart Brand himself played a part in the release of an image of planet earth, when in 1966 he wondered why, despite NASA's space programme, there were no images of the planet from space. He thought 'seeing an image of Earth from space would change a lot of things' (cited in Kirk 2007: 40). In 1966, Brand had small badges printed with the rhetorical question: 'Why haven't we seen a photograph of the whole Earth yet?'. Brand used one of the first colour images of the earth as a disc taken from a weather and communication satellite in 1967 for the cover of the first *Whole Earth Catalog*.

The *Whole Earth Catalog* is a strange potpourri of product and book reviews designed to empower the individual. The opening paragraph of the first edition states that the purpose is to give power to the individual 'to conduct his own education, find his own inspiration, shape his own environment and share his adventure with whoever is interested. Tools that aid this process are sought and promoted by the *Whole Earth Catalog* (1968). Steve Jobs once described the *Whole Earth Catalog* as 'Google in paperback form, thirty-five years before Google came along' (cited in Wiener 2018).

Andrew G Kirk observes:

> The ecological sensibility conveyed through the *Whole Earth* captured a new alchemy of environmental concern, small-scale technological enthusiasm, design research, alternative lifestyles, and business savvy that created a model of environmental advocacy.
>
> (Kirk 2007: 2)

The *Whole Earth Catalog* optimistic, can-do discourse acted as a counterpoint to the prevailing eco-apocalyptic discourses of the time, exemplified by the San Francisco Diggers and Barbara Ward's *Spaceship Earth*. The *Whole Earth Catalog* was also a practical utopian discourse as it critiqued contemporary industrial and consumer society, imagined a better way of being and suggested practical solutions to environmental and social problems.

Popularising Environmental Science

The dominant pessimistic and indeed apocalyptic tone of environmental concern can also be identified in two significant scientific accounts that became extremely influential – Rachel Carson's *Silent Spring* published in 1962 and the *Limits to Growth* published a decade later in 1972. Both texts can be considered as a popular science, as they were founded in scientific research, but were intended to address a much wider audience than the scientific community. Both also sold well and attracted controversy.

Rachel Carson's book *Silent Spring* sold over half a million copies and was on bestseller lists for 31 weeks. John Urry (2011: 94) suggests that Carson's book 'initiated new kinds of environmental arguments and practices, as opposed to those more concerned with landscape and countryside conservation'. *Silent Spring* is a significant scientific analysis of the potential devastation wreaked by chemicals, particularly the insecticide DDT on life. Carson was a marine biologist, but it is obvious that she intended the readership of the book to extend well beyond the scientific community. She begins her argument with a poetic description of a rural idyll that is silenced by the poisonous effects of unrestrained uses of chemical pesticides and fertilisers. While the small rural community that she represents as blighted by the toxic chemicals of modern agriculture did not exist, nor was there any single place that had experienced all the calamities she vividly describes, yet 'every one of these disasters has actually happened somewhere and some communities have already suffered a substantial number' (Carson 2000: 22). M. Jimmie Killingsworth and Jacqueline

Palmer suggest that *Silent Spring* 'possessed a rhetorical power unmatched in its day'. This rhetorical power 'tapped into ... the public's growing uneasiness with science' (Killingworth and Palmer 1996: 27). Paradoxically, Carson's argument is grounded in science and contributed to public anxiety about science.

Tim Radford in a short retrospective article for *The Guardian* wrote:

> If you had to choose one text by one person as the cornerstone of the conservation movement, the signal for politically savvy environmental activism, and the beacon of worldwide lay awareness of ecological systems, *Silent Spring* would be most people's clear choice.
>
> (Radford 2011)

The Limits to Growth was a very different style of publication to *Silent Spring*. It was sponsored by The Club of Rome – an informally constituted group of scientists, educators and economists – and produced by an international team of scientists coordinated by the prestigious Massachusetts Institute of Technology (MIT). The research was the first computer modelling of global trends. The *Limits to Growth* report concluded:

> If the present trends in world population, industrialization, pollution, food production and resource depletion continue unchanged, the limits to growth on this planet will be reached sometime within the next one hundred years.

They also suggested:

> It is possible to alter the growth trends and establish a condition of ecological and economic stability that is sustainable far into the future.
>
> (Meadows et al 1972: 23–4)

Finally, they argue that the sooner we start working to turn around the trends that are hurtling us towards the limits of growth and redirect humanity's path to a more sustainable future, the greater will be the chance of success.

The authors of *The Limits to Growth* indicate that the implications of their computer modelling reach 'far beyond the proper domain of a purely scientific document. They must be debated by a wider community than that of scientists alone' (Meadows et al 1972: 23). Consequently, the report was written in a style that was intended to be accessible to the general public. Nonetheless, the style of the report tends to emphasise scientific rigour and objectivity. *The Limits to Growth* aims to demonstrate that the experts have an insight into the state of the world. While the authors admit that there are several shortcomings to their modelling, they assert that 'we do not expect our broad conclusions to be substantially altered by further revisions' (Meadows et al 1972: 22). In many ways, the rhetorical style of CAT's publications, particularly in the later years, has tended towards a style that more resembles *The Limits to Growth*.

While *Silent Spring* and *Limits to Growth* emphasised the problem, and could be considered apocalyptic in tone, there was also a more positive discourse that was more utopian in style. We have already encountered this positive approach in the *Whole Earth Catalog*, which is a resource for potential solutions to environmental issues and the anomie felt by the counterculture caused by urbanisation, industrialisation, capitalism and consumerism. However, there was arguably an even more important text that fed into the utopian stream of AT, namely, E.F. Schumacher's *Small is Beautiful* first published in 1973. Schumacher (1911–1977) was a Swiss economist, who moved to England in 1936. Radkau describes Schumacher as 'one of those rather genial prophets who did not announce the coming apocalypse but aimed to offer hope and pragmatic solutions' (Radkau 2014: 190). *Small is Beautiful* is a collection of essays that challenge a central idea of modern economics that 'more is better'. Schumacher questions the idea that large-scale industrialisation, unlimited economic growth and unbridled wealth are key indicators of success. Schumacher's argument is about scale. He states that 'today we suffer from an almost universal idolatry of giantism. It is therefore necessary to insist on the virtues of smallness' (Schumacher 2011: 49). For example, Schumacher argued that large-scale industry was inherently destructive to the well-being of both people and the environment. The solution according to Schumacher is:

> The evolution of small-scale technology, relatively non-violent technology, 'technology with a human face', so that people have a chance to enjoy themselves while they are working, instead of working solely for their pay packet.
>
> (Schumacher 2011: 9)

He argues that 'small-scale operations, no matter how numerous, are always less likely to be harmful to the natural environment than large-scale ones' (Schumacher 2011: 22).

In the introduction to the Vintage edition, the British environmentalist Jonathon Porritt writes: '"Small is Beautiful"! That deceptively simple notion still resonates very powerfully throughout the Green Movement' (in Schumacher 2011: x). More specifically, Witold Rybczynski (1991: 6) observes that 'E.F. Schumacher was undoubtedly the motive force behind the AT Movement'.

These three texts – *Silent Spring, The Limits to Growth* and *Small is Beautiful* – can be considered as the foundational texts to the modern environmental movement and specifically inform the AT perspective.[2] Almost all of the early pioneers of CAT, including the founder Gerard Morgan-Grenville, state that they had been inspired by *Silent Spring* and *The Limits to Growth* and/or *Small is Beautiful*. CAT is also part of the Schumacher Circle – 'a family of organisations in the UK that have all been inspired, one way or another, by E.F. Schumacher' (The Schumacher Institute 2021).

While we are looking at the texts that have played a significant role in informing AT, it is necessary to quickly mention two UK magazines *Undercurrents* and *Resurgence*. *Undercurrents* described itself as 'a magazine of radical science and alternative technology'. *Undercurrents* was published several times per year between 1972 and 1984 and contained a diverse range of articles ranging from discussions on capitalism to cheesemaking, from the Gaia hypothesis to Living without a Television. *Undercurrents* was the arena, pre-internet, where ideas were shared and disseminated about the philosophy, ideology and practice of AT. *Resurgence* was started in 1966 but merged with *The Ecologist* in 2012. *Resurgence* tends to focus more on the spiritual and philosophical aspects of environmentalism. Both E.F. Schumacher and James Lovelock as well as other renowned authors have contributed articles to *Resurgence*.

It was out of this milieu – the countercultural challenge to the dominant ideology and cultural mores of the developed world; the growing consciousness of the earth as a finite bounded planet; a growing ambivalence about the consequences of industrialisation and technology; and the popularisation of scientific concerns about the environment – that modern environmentalism emerged. AT is a particular manifestation of modern environmentalism, and it is therefore necessary to consider the connotations of the term.

Alternative Technology: What Is in a Term?

In many ways, the interplay between the countercultural and the scientific are captured in the term 'alternative technology' (AT). This term is attributed to Peter Harper, who joined CAT in the early 1980s. Harper (2014) states that the term AT came about when he required a name for a conference that he was organising in 1972. Harper (2014) suggests that the first published use of the term appeared in a short guide to AT that he jointly wrote with Björn Erikson, which appeared in the third (Autumn/Winter) edition of *Undercurrents* in 1972. Harper (2014: 2) acknowledges that the term 'was intended to bridge the gulf between the techno-manic mainstream and the charming but ineffectual anti-technological counterculture'.

> AT is a sort of oxymoron, you know, because the alternative and technology tend to sit in different parts of the brain. So, if you put the two together there is a bit of a frisson. Alternative tended to mean a complete raggle-taggle, no organisation and a completely anarchic situation. Technology offered a bit of a spine stiffening element.
>
> (Peter Harper, interview, November 2020)

Harper (2014: 2), suggests that the concept of AT includes not only the mechanical hardware, such as windmills, but also things such as gardening and diet, and he observes that 'it is hard to see any limits to its scope'.

AT was not the only term that was used in the early modern green movement in the UK. Peter Harper and Björn Erikson in their guide to AT in the third edition of *Undercurrents* list a number of other terms such as New Alchemy,[3] Ecologically-Based Technology, Liberatory Technology[4] and People's Technology. Although Harper and Erikson indicate that each of these terms implies a slightly different focus, these are 'technologies that are designed to satisfy directly human needs' rather than 'productivity and profits' (Harper and Erikson 1972: 23). This indicates that AT, unlike large-scale industrial technology, is not intended to serve the capitalist system. Consequently, while CAT is not overtly political in the same way as green political parties, it has always implicitly and frequently explicitly challenged unrestrained consumerism and the central idea of capitalism that economic growth is inherently good. There are two reasons why CAT has never been overtly political. The first and perhaps most important reason is that it is crucial that the message of AT is as inclusive as possible and does not solely resonate with one part of the political spectrum. Second, CAT is a registered charity, and in the UK, there are very clearly defined legal limitations to the political engagement that charities can undertake. While a charity may campaign for a change in the law, 'it cannot exist for a political purpose, which is any purpose directed at furthering the interests of any political party' (Charity Commission for England and Wales 2008).

Peter Harper himself used the term 'Soft Technology' for a UNESCO report published in 1972. In 1976, Harper and Godfrey Boyle edited a book with the title *Radical Technology*. In the preface they indicate, 'This is a book about technologies that could help create a less oppressive and more fulfilling society' (Boyle and Harper 1976: 5). This clearly alludes to the utopian aspects of the early green movement. Indeed, David Dickson uses the term 'Utopian Technology', which is 'based on a sense of co-operation, rather than competition and domination between man and nature' (Dickson 1974: 98). Other terms that were used at the time were the linked terms 'Intermediate Technology' (IT) and 'Appropriate Technology'. Intermediate Technology, a term coined by E.F Schumacher, refers to technology that is appropriate for developing economies that are capital poor but labour rich. This technology, according to Schumacher, is intermediate in a number of senses. First in the economic sense in that IT would be less capital intensive than the technology of modern industry but would be more capital intensive than indigenous technologies. Second, IT would be more productive than indigenous technologies but less productive than industrial technologies. Third, IT would be less complex than modern industrial production but more complex than indigenous technologies. In this sense, IT could be more easily understood and maintained at a local level (Schumacher 2011: 148–50). Schumacher founded the Intermediary Technology Development Group (ITDG)[5] 'to assist in the collection and dissemination of data on simple low-cost technologies' (Dunn 1978: 4). ITDG's agenda was primarily to assist the developing world. The term 'Appropriate Technology' was also 'applied to situations in developing countries and was derived from Schumacher's earlier notion of 'Intermediate Technology' (Harper 2016: 429).

What unifies the majority of these is the term 'technology', which from the 19th century has been widely understood as the practical application of science (Williams 1988: 315). Technology is also understood in instrumentalist terms as a means to an end. It is also frequently associated with machines that are regarded as enabling the achievement of specific ends, easily, quickly and comfortably. This conception of technology is clearly linked to ideas such as progress, efficiency and convenience (Slack and Wise 2015). As well as mitigating the perceived dissolute hedonism of the alternative movement – to add 'a spine-stiffening element' to use Harper's description – the use of the term technology is intended to counter the idea that a sustainable lifestyle entails regressing to a life of primitive hardship. AT is neither a return to the stone age devoid of comfort and convenience, nor a romantic utopian pipe dream, but is best described as practical utopianism.

The use of the term technology can easily be misperceived, as it seems to emphasise the mechanical hardware of renewable energy – wind and water turbines, and solar panels. It also seems to suggest the idea that there is a simple technological fix to environmental collapse. It is simply a matter of replacing non-renewable and polluting technologies with renewable and clean technologies. Consequently, the qualifiers, such as Alternative, Appropriate, Soft, Radical, are intended to indicate the social, political and cultural implications of technology. These qualifiers suggest a dualistic view of technology – for example, there are appropriate or inappropriate forms of technologies, soft or hard manifestations of technology. More than any of the other qualifiers, 'Alternative' signifies a very Manichean view of technology. The term 'alternative' suggests a radical distinction between the technology of modern industrial society, which is construed as large scale, centralised, alienating, exploitative and polluting, and AT, which is envisaged as small scale, human, non-exploitative and more in tune with nature.

This Manichean view of technology is clearly identifiable in David Dickson's 1974 publication *Alternative Technology*. In the first part of this book, Dickson outlines the problems of what he calls 'contemporary technology', such as 'environmental pollution' and 'the depletion of the world's non-renewable natural resources' (Dickson 1974: 17). The issue is, according to Dickson, not simply a practical one of developing and adopting different types of technology but an ideological one. Dickson argues that an AT is based on an ideology of 'ecological and social considerations' rather than the industrial ideology of 'economic efficiency' (Dickson 1974: 101). Consequently, AT does not simply propose adopting different forms of technology – switching from fossil fuels to renewable energy sources – but directly opposes the prevailing ideology and mores of contemporary society. This oppositional conception of AT is clearly articulated in the very first edition of *Undercurrents*. In their manifesto, the editors state:

> Technology, too, while still masquerading as mankind's greatest emancipator, is increasingly becoming the instrument of our enslavement. Though it continues to be regarded as simply the application of scientific rationalism to the satisfaction of human needs, technology in practice is the

means whereby the unjust economy and power structures of our industrial society is kept intact and entrenched ... .

We also believe that technology can be reoriented to serve not economies and governments but individual human beings.

(Undercurrents 1972: 2)

As indicated above, the term alternative in the late 1960s and early 1970s has associations with the counterculture. Throughout its history, CAT has struggled with the perception of a hippie image by many outsiders. It remains a moot point, whether the two terms 'alternative' and 'technology' successfully modified each other or whether alternative connoted too radical an approach for the mainstream and technology signified too hard an approach for the counterculture.

There is a considerable debate at CAT today, whether AT is really an apposite term in the current context. Some current members take issue with the term alternative, some with technology and some with both. The general objection to the term is that CAT is no longer alternative as environmentalism is very much mainstream now. This might simply be because the hardware of AT is no longer alternative.

> The technology is not alternative anymore. If you drive around any suburban housing estate, how many solar panels do you see? Electric cars are increasingly prevalent. Wind turbines are fairly ubiquitous. In the early days it was about developing new technology. CAT was as close as you got to the cutting edge of developing these new technologies. We do not do that now because the technology has gone mainstream. Siemens or somebody have probably got a billion dollars' worth of laboratory facility to work in. Our emphasis has changed more from developing technology to explaining technology. Explaining to people how we get to net zero carbon.
>
> (Peter T, interview April 2021)

Anna, who is the manager of the ZCB Innovation Lab at CAT (see Chapter 5), suggests that the concept of technology has to be reframed so that it no longer simply refers to mechanical things that are useful.

> What I'm trying to encourage is a narrative around social technologies. So I think this whole innovation lab and systems thinking approach is like a social technology where people need to think differently, work differently, act differently, interact differently. So I really think that could be a better sales pitch for CAT is to say we have got the physical technology. Now we need to address the social technology.
>
> (Anna, interview July 2021)

Here Anna alludes to a much wider understanding that technology is much more than useful things for achieving particular ends. The term technology derives from the Greek *techne*, which is generally translated as art, craft, technique and skill. Martin Heidegger suggests that '*technē* belongs to bringing forth, to *poiēsis;* it is something poietic'. The Greek term *poiēsis*, although it signifies 'to make', means something much more than making a physical thing. In Heidegger's thinking, 'it is as revealing, and not as manufacturing, that *technē* is a bringing-forth' (Heidegger 2013: 13). In this sense, technology, in Anna's view, is the bringing-forth of a better society. If the physical technology exists to turn around the environmentally calamitous effects of human activity, the question that we have to address, as a matter of utmost urgency, is why we are not bringing forth a more sustainable society.

Some of the current staff are sceptical about the use of the term 'alternative'. They accept that while CAT might have been alternative in the past, it no longer is and in the current context it might be counterproductive to be identified as being alternative. However, others suggest that the designation 'alternative' is still useful.

> Alternative I still think it works because although CAT is more mainstream and it is not so experimental anymore, it is still an alternative to the norm for a lot of people. So I think another way of looking is that alternative also means choices. And what we are actually doing is showing people choices.
> (John C, interview September 2020)

AT can be seen as the choices that we make to bring about a sustainable society. However, it is generally agreed that, although in many ways it has moved from margins to the mainstream and that the name is rather anachronistic, CAT because of its history has a sort of brand recognition.

CAT: The Beginning

The term 'alternative technology' was selected by the founder Gerard Morgan-Grenville for his new embryonic centre amongst this plethora of terms. In many ways, this choice perpetuated the use of the term 'alternative technology' in the UK. The term 'alternative technology' was only used in one other context, which was the Alternative Technology Group[6] set up at the Open University by Godfrey Boyle, the editor of *Undercurrents*.

Most people know and associate the term AT with renewable energy, and this is in part because both CAT, at least in the early days, and Godfrey Boyle's AT unit tended to emphasise the more technological aspects. Indeed, sometimes CAT was simply referred to as the Technology Centre. Peter Harper and Godfrey Boyle, neither of whom were involved in CAT in the very early pioneering days, reminisce in their conversation for the oral history project, that while they had been talking and publishing about AT in the early 1970s, they were somewhat taken aback when they heard about the embryonic centre being started 'by some mysterious people in a quarry in Wales'. Though when he visited, Boyle said that he was impressed by what they were doing.[7]

The Centre was originally called the National Centre for Alternative Technology, but soon changed its name to simply the Centre for Alternative Technology, as it was ambiguous whether national signified Wales or the UK more broadly. The April 2013 edition of the Radio 4 series *The Reunion*, which brought together a group of the early pioneers, described CAT as 'the radical community that launched the Green Movement in Britain from a disused slate quarry in Machynlleth, Wales'. Sue MacGregor in her introduction to the show identified CAT as 'One of Europe's leading showcases for sustainable living' (BBC Sounds 2013).

CAT was founded by Gerard Morgan-Grenville (1931–2009). In the context of the early environmental movement, Morgan-Grenville seems an unlikely individual to be the founder of CAT, being neither from a scientific background nor countercultural. Morgan-Grenville was from the privileged classes, his great grandfather was the last Duke of Buckingham. He was educated at Eton, and then did his military service in 1949, becoming the aide-de-camp (ADC) to General Festing. After being demobilised, he and his brother set up a fairly successful trading business. However, Morgan-Grenville observes in his autobiography that the rationale for establishing the business was purely financial and that while his brother found managing a company a rewarding career, he did not and therefore he decided to leave the business (Morgan-Grenville 2001: 134–5). Morgan-Grenville took up painting and, on an impulse, purchased a barge in the Netherlands with the idea of it becoming a floating studio. He eventually started taking paying passengers as a way of financing the barge, and this became a thriving business and was a pioneer of leisure cruises on the rivers and canals of France.

In his autobiography, Morgan-Grenville indicates that there were three things that inspired him to found CAT. These were reading Rachel Carson's *Silent Spring* and journalists such as Gerald Leach who were beginning to report on environmental concerns; a serendipitous meeting with Diana Brass whom he characterises as 'a Green Fundamentalist' (Morgan-Grenville 2001:154–5); and a journey to the USA with Diana Brass where they encountered the counterculture. Returning from the USA, both Brass and Morgan-Grenville felt compelled to do something about the threat that they perceived to the environment. Morgan-Grenville was very ambivalent about the counterculture. On one level, he admired the hippies for their attempt to find an alternative lifestyle, but he was also rather critical of what he perceived as their ill-disciplined approach to life.

> They were mostly muddled and disorganised, fragile. Most failed, but the important thing was that they had opted out of the mainstream life in order to find a way of living which respected the environment in which they lived. They also rejected Authority in principle. I found myself in sympathy with both aims.
>
> (Morgan-Grenville cited in Centre for Alternative
> Technology 1995: 4)

According to Roderick James, the first director of CAT, there were two prime inspirations in the USA – The New Alchemy Institute[8] and the Farallones Institute. The New Alchemy Institute in Cape Cod started in 1969 was formally incorporated in 1970 by John Todd, his wife Nancy and William McLarney. New Alchemy was similar to CAT in trying to find practical solutions to environmental issues. John Todd told the *New York Times* 'I was tired of ringing alarm bells all the time. I want constructive alternatives' (cited in Kirk 2007: 144). Both Todd and McLarney were marine biologists. Todd stated that the aim of New Alchemy was both scientific and social. On the scientific level, New Alchemy researched alternative forms of agriculture and power generation that would not deplete the earth's resources. Todd explains:

> On a social level we seek to aid and foster the development of decentralized communities where people can live in a manner consistent with our ecological principles.
>
> (cited in Eldred 1989: 4)

The Farallones Institute was founded by the architect Sim Van der Ryn, who was integral for developing the idea of sustainable architecture, which is a core aspect of AT (see Pursell 2009). Van der Ryn also designed the Integral Urban House in Berkeley, California, which was an experiment in self-reliant living in an urban environment.

Diana Brass's and Gerard Morgan-Grenville's skills and characters were mutually complimentary. Morgan-Grenville was the ideas person who with his various connections in the establishment made things happen in the background. His establishment status in many ways ensured that CAT, particularly in the early days, had a certain credibility that enabled it to survive. Brass, who has acquired an almost mythical status with her 'can-do' approach, is represented as the unifying heart of early CAT.

Morgan-Grenville indicates that his vision was to set up a centre:

> Where people, ordinary passers-by, might readily perceive the disastrous course on which our civilisation was set and be shown things they, *anyone,* might do to reduce their impact on the environment.
>
> (Morgan-Grenville 2001: 158)

There were two central planks to this vision: Firstly, to raise awareness of the harmful impact of human activity on the environment. Second, and more importantly, to offer some practical solutions to the perceived environmental disaster. It is this second agenda that made CAT unique and different from campaigning groups such as FoE and Greenpeace, which were founded at around the same time. This agenda of finding practical solutions had three intertwined aspects – experimentation and finding what worked; living a sustainable lifestyle and all that might entail; and education and outreach.

First, a suitable location, a formal identity and funding had to be found. The latter two were fairly straightforward. Morgan-Grenville founded a charity called the Society for Environmental Improvement in January 1973[9] and managed to obtain the endorsement of a number of establishment figures including Roy Jenkins, the then Home Secretary, Lord Robens, the renowned ornithologist Peter Scott, and ironically Derek Ezra who was at the time the Chair of the National Coal Board. In an interview for *Undercurrents*, Morgan-Grenville was asked about the number of eminent names on the letterhead; he replied that it was a very deliberate policy to seek the endorsement from people who 'head industry, government, large organisations' (Morgan-Grenville 1974: 13). This was both a really canny strategy, but also came under criticism from the countercultural branch of the AT Movement. As Adrian Smith (2005) argues, AT, as the name implies, is inherently antithetical to industrial capitalism, advocating small-scale and decentralised enterprises:

> The development of AT presented activists with a dilemma. Should they compromise in order to appeal with less enthusiastic business and government interests, or remain true to their original, radical critique?
>
> (Smith 2005: 109)

For many, the involvement, however minimal, of establishment figures was a compromise too far. However, Morgan-Grenville's establishment identity and connections gave CAT a certain status and support it might well not have garnered in the very early days. Nonetheless, Morgan-Grenville was also radically critical of the prevailing attitudes of consumer culture, and the idea that perpetual economic growth is inherently good for individuals and society. In an early document that Morgan-Grenville wrote in 1973, just after the acquisition of the site but prior to anyone moving there, he sets out his vision. It is not clear how widely, if at all, this document was circulated. As this document was simply typed and as far as I know there was only a single copy, it is unclear whether it was simply a way for Morgan-Grenville to clarify his thoughts, whether it was intended for potential supporters or to shape the direction of the project.[10] Although it perhaps took CAT longer to develop in the way that Morgan-Grenville envisaged, it did substantially manifest in the way that he delineated in this position document.

In this document, Morgan-Grenville first defines AT as 'the techniques which have minimal adverse consequences to the environment'. However, he also states that addressing environmental issues is not simply a technological fix but also requires a philosophical reorientation.

> [AT] is being seen not only as a technical solution to current problems, but as a philosophical alternative to man's obsession with material expansion and its inevitable association with economic and social chaos.
>
> (Morgan-Grenville 1973)

In an allusion to *The Limits to Growth* and a specific reference to the oil crisis, Morgan-Grenville indicates that one of the prime issues are finite resources. However, this is not simply a matter of exhausting oil and other resources. He states that 'it is morally indefensible to consume the strictly finite resources of the earth'. In other words, Morgan-Grenville makes a philosophical and moral argument. As well as *The Limits to Growth*, Morgan-Grenville clearly alludes to Schumacher's ideas when he states that: 'there is also a growing awareness that the administrative and manufacturing centralisation necessary to sustain rapid material expansion is dehumanising people' (Morgan-Grenville 1973). Centralisation is concomitant with the prevailing ethos of industrial capitalism and the advocacy of economic growth. Larger corporations, through the economy of scale, have the potential to generate greater profits and cheaper consumer goods. However, Morgan-Grenville suggests that this centralisation is inherently detrimental to human flourishing. This clearly alludes to Schumacher's central thesis that small-scale technology is both relatively non-violent and has 'a human face'. This perspective becomes an important aspect of CAT's ethos, particularly in the early years.

Having described the problem, Morgan-Grenville outlines his perceptions for his embryonic project, which was to 'establish a national centre to demonstrate that the average needs of a domestic situation can be met substantially by alternative technologies appropriate to the services required' (Morgan-Grenville 1973). The focus of the proposed centre would be on energy, recycling, building, food and transport. Morgan-Grenville saw the centre as being an educational resource, a place for experimental research and demonstration of AT. He also saw the community as integral to the project.

The Quarry: 'A Happy Combination of Vision and Place'

To fund the nascent project, Morgan-Grenville asked his elder half-brother for £20,000 seed money (Morgan-Grenville 2001: 159). However, finding a suitable site proved more challenging. Morgan-Grenville sent Steve Boulter, an American PhD student, around the UK in search of a site. Eventually, Boulter stumbled across a disused slate quarry just outside of Machynlleth in mid-Wales. Hence the site is often simply referred to as The Quarry. The site, hidden from view, was mostly piles of massive slabs of slate and was overgrown with brambles. There was a row of three semi-derelict cottages and two large, ruined engine sheds. Steve Boulter in an appeal for volunteers in *Undercurrents* observed that this disused slate quarry:

> Had just the combination of wind, water, sun, space and surroundings that we are after … The site came equipped with all the essentials including 6,000 square feet of ancestral slate cutting sheds, three cottages, two streams and reservoirs, assorted tunnels, a bit of railway and several thousand tons of slate of assorted sizes and shapes.
>
> (Boulter 1974: 4)

As human geographers and ethnographers have observed, place has a character that can shape activity, experiences and meaning. Conversely, human activities and narratives of meaning also construct what has been called the *genius loci* – the sense that a place is unique and distinguishable from all other places. The location and physical aspects of the rural Welsh environment in many ways structured the way in which CAT developed, but at the same time, the everyday activities at this disused slate quarry created CAT as a unique place. Many people, in both the oral histories and in the interviews that I conducted, commented on the special character of the place. This specialness of the site is articulated in terms of the history of CAT, the reclaiming of the industrial landscape and its natural beauty.

The landowner John Beaumont, who had coincidentally been at Eton at the same time as Morgan-Grenville, agreed a peppercorn rent of one shilling (5 pence) per annum on a 100-year lease. Again Morgan-Grenville's social status was a significant factor. Morgan-Grenville observed:

> Where better than this derelict but beautiful site to pioneer a way of life which could be lived without using up the capital resources of planet Earth.
>
> (Morgan-Grenville 2001: 162)

In the position document, Morgan-Grenville stated that 'part of the A.T. "philosophy" is decentralisation. For this reason a site away from a major town is preferable'. He also noted that 'there is a good rail service to the town of Machynlleth'. (Morgan-Grenville 1973).

Peter Harper describes CAT as a 'happy combination of a vision and a place' (Centre for Alternative Technology 1995: 4). The location proved both a blessing and a curse. The location was beneficial as it was fairly isolated, it enabled the development of the community and it is located in an extremely beautiful part of rural mid-Wales. However, the geographical remoteness also seemed to emphasise the remoteness of the ideas. It was common to hear the criticism that it was fine to lead a green lifestyle in remote mid-Wales, but that this alternative lifestyle was not relevant in an urban context and was frequently represented as being alien to most people's lives. Mark Mathews who was employed as a sort of site manager jointly with his wife Mary at the embryonic CAT indicated that 'in those days people thought we were daft' (Mark Mathews, Oral History).

Pioneering Days

Soon a number of volunteers started to arrive. Morgan-Grenville (2001: 166) describes many of these early pioneers as 'an uneasy mix of anarchists, communards and neo-primitivists'. As indicated above, Morgan-Grenville had a very clear vision of what he hoped to achieve but in many ways this was incredibly ambitious, particularly given the remoteness of the site, the very derelict nature of the few remaining buildings, the Welsh weather and the sometimes rather limited skills of some of the early pioneers. Roderick James, who became CAT's first Director, characterised these very early days in his interview for the oral history project as an 'extraordinary clash of old-Etonian and hippie, which was never fully resolved'.

Morgan-Grenville stayed mostly in the background making connections, hustling for supplies and occasionally sending imperious directives on embossed paper. Nonetheless, the overwhelming consensus was that Morgan-Grenville was incredibly charming and energetic. Jill Whitehead, a very early volunteer, indicated that all that really mattered to Morgan-Grenville was realising the objectives that he set, and that personally for him it did not really matter if he had his name put on the project. However, being on-site was incredibly tough in the first year. Peter Harper (2015: 3) describes the early days as 'reminiscent of a refugee camp or post-disaster situation'. Work started on renovating the cottages in order to make them habitable. One of the engine sheds that was still partially roofed with corrugated iron was patched up and used as a communal area.[11] A typical entry from a diary that was kept during this pioneering period reads:

> Heavy rain all day. The rain has done some damage to the road requiring two of us to ditch and fill. The front of the cottage has been glazed by Pat. It is not possible to floor the front room yet, as no polystyrene is available in Machynlleth. Another seven people have arrived, so we are now twelve.
> (Friday 8 February 1974)[12]

Diana Brass mostly stayed on-site and seemed to have been the core spirit in those early pioneering days. Morgan-Grenville (2001: 166) describes her as 'unpaid and asking no thanks she provided the motherly glue which kept everyone together'.

> Diana as usual has been marvellous, keeping us well fed and never ceasing to work. She cleared the rotting floor from the engine shed and reorganised the cottage in a much tidier way.
> (Diary entry Sunday 19 February 1974)[13]

There were a number of passionate, and sometimes heated debates as to the direction that CAT should take. There was clearly a clash between Morgan-Grenville's grandiose vision and some of the more cautious and pragmatic views. Tony Williams,[14] who was appointed as the first site manager in February 1974, after a meeting with Morgan-Grenville and others wrote:

> I am disturbed by the proposed size of the project, and that we seem to be split between those that think big, and those who are more down to earth … Steve[15] is still hung up on his exhibition hall, although no one has any idea what it might be.[16]

There was also a debate about how to preserve the natural beauty of the site. Tony Williams in the diary expresses anger at Steve for removing some vegetation. In a later entry he observes: 'it is going to be very difficult to preserve much of the existing ecology here'.[17]

Consolidation

The turning point, that ensured CAT's survival was the appointment by Morgan-Grenville of Roderick James in 1975 as the first Director on a small salary. One of James's first and most canny decisions was to place an advertisement in *The Times* asking: 'Have you a month to spare?'. As James indicates on *The Reunion* programme on BBC Radio 4, the *Times* initially rejected the advert as they perceived that CAT was 'an anarchical organisation'. However, with Morgan-Grenville's establishment credentials, *The Times* did eventually publish the advertisement, and this led to a number of people arriving at the Quarry with the necessary skills to drive the project forward.

Perhaps the most important of these new arrivals was Bob Todd, who became the technical director.

> I have to say that if Bob had not turned up at the crucial time CAT would not be there. It was only Bob really that gave CAT its credibility.
>
> (Peter Harper, Oral History)

If Morgan-Grenville gave CAT a certain establishment credibility, then Todd provided its scientific authority.[18] Bob Todd has a PhD in electrical engineering. He had worked on electronic controls for medical equipment but was teaching at Southampton University when he heard about CAT on a programme about the Ideal Home Exhibition that was donating a kit house to the Centre.[19] Todd had a nascent interest in the environmental movement having read *The Limits to Growth* and had started to consider how renewable energy systems might function in the UK. Todd is also incredibly practical – he was a 'hands-on' engineer as well as an academic and had a set of skills that was largely lacking at the time. In 1974 at a time when volunteers were still struggling to make the site habitable, Todd went to have a look at what was happening.

> I immediately got drawn into putting in a small hydro-electric turbine. There had been a problem with [installing] that because there was a lack of expertise. I committed myself to building an electronic controller for it.
>
> (Bob Todd, Oral History)

Todd was persuaded by Morgan-Grenville to move to Wales, and he worked at CAT for the next two decades. Bob Todd is very typical of those who advocated a more environmentally sustainable approach in that he saw a clear link between sustainability, social justice and an egalitarian approach to life. Sustainability in this view is not simply a technical fix for the destructive trends of human activity outlined in *The Limits to Growth*, but for Todd it also has an ethical dimension.

> In my head the two [sustainability and egalitarianism] are coupled in the sense that caring about sustainability is tied up with ethical considerations of fair sharing around the world.
>
> (Bob Todd, Oral History)

Community or Outreach?

Perhaps the most significant debate in the early days of CAT was between those whose main focus was on the formation of a self-sufficient intentional community, and those who were convinced of the necessity to educate the wider public about more environmentally friendly lifestyles.

> There was always a tension between those who wanted to raise a drawbridge to the outside world and those who believed that what they were doing was primarily to serve others.
>
> (Richard St George cited in Centre for Alternative Technology 1995: 9)

It is clear that Morgan-Grenville intended CAT to be relevant to a wider public. In an interview for *Undercurrents,* he clearly states:

> I think a very important point to realise is that we have got an external interest here at the centre, whereas most communities are internal – they're interested in their own survival, doing their own thing. Of course we're interested in that too, but we are also here to serve people outside.
>
> (Morgan-Grenville 1974: 14)

This debate between the more inward-looking faction that was more intent on founding a viable intentional community and those more concerned with outreach was also implicated in a further debate between what might be called the countercultural ethos, which believed that AT necessitated rejecting all aspects of the establishment, and the more pragmatic ethos, which perceived the necessity of engagement. This clash of vision is epitomised by the visit to the embryonic Centre in October 1974 by the Duke of Edinburgh. *Undercurrents* (Issue 8) published an edition with an image on the front cover of a windmill on top of *Buckingham* Palace with the tag 'Royal Fillip for Alternative Technology'. The short report about the royal visit suggested that 'Alternative Technology until recently the almost exclusive province of cranky eco-eccentrics has become *respectable* with a rapidity that has taken most AT freaks' breath away'

(Boyle 1974: 1. Emphasis in original). Godfrey Boyle indicated that the report was intended 'to be sardonic' but that it made him quite unpopular at CAT for a while (Godfrey Boyle, Oral History).

Morgan-Grenville (2001: 173) himself notes that 'on the eve of The Visit there were animated discussions. The anarchists had taken themselves off in a spirit of extravagant disapproval'. However, the CAT diary also suggests that a letter was written to a couple of the more radical volunteers suggesting that they should absent themselves on the day of the Royal visit. Morgan-Grenville recalls that in the animated discussion prior to the Duke of Edinburgh's visit, the radicals suggested that the Centre was not sufficiently 'alternative', to which Morgan-Grenville replied that 'the brief was a little broader, trying to establish a bridge within the system'.[20] This highlights a major debate within the green movement more generally. If the assessment is that environmental degradation is caused by human activity and consequently 'business as usual' is no longer possible, the dilemma is: should one, as Morgan-Grenville suggested, try to change the system from within or, as his detractors argued, set up radical alternatives outside the system? Those who propose a radical alternative argue that working within the system entails compromises that derails the environmental agenda. Those who suggest that the only effective strategy is to work within the system perceived that being outside the system cannot bring about the necessary changes.

In a meeting between Roderick James and Morgan Grenville, it was determined that outreach and education must be part of the project; James suggests that during this meeting 'the blueprint for the Quarry' was laid, and that it would be 'essentially a visitor centre' with demonstrations of AT. Although this apparently met with some resistance, this established the direction that CAT was to take (Roderick James, Oral History). In July 1975, CAT opened as a visitor centre. Gerard Morgan-Grenville (2001: 175) indicates that by the end of the year there had been 15,000 visitors to CAT.

It was in these early days that the core principle of consensus decision making was established. Weekly meetings of everyone involved were held, which Roderick James described as 'incredibly successful, but time consuming' (Roderick James, Oral History). In many ways, it was this levelling of the organisational and decision-making structures enabling people to feel actively involved that partly accounts for CAT's success. While the actual mechanisms for participation in decision making were modified over the years, the core principle of consensus decision making remained in place until 2010 when a more hierarchical management system was instituted.

While CAT would never have got off the ground without Gerard Morgan-Grenville's vision, connections and work in the background, he was regarded as being too autocratic. According to Roderick James, Morgan-Grenville would send memoranda on embossed paper indicating that standards and progress were not what he expected. This led to a major row about the control of the

Quarry. Some of the people who were living and working full time at the Quarry came to resent Morgan-Grenville, who seemed to be leading a very comfortable life off-site, sweeping in and telling people what to do. Roderick James describes a meeting between Morgan-Grenville and the early pioneers as 'very stormy and quite painful' (Roderick James, Oral History). After this confrontation, Morgan-Grenville graciously stepped further in the background, although he remained the Chair of the Board of Trustees. Those who were working full time took over the running of the Centre. With a core of highly skilled and dedicated staff, the Centre began to thrive as a community, a visitor attraction and an educational centre. However, these different aspects resulted in CAT developing in a highly convoluted way in which these various strands of the project interweaved.

> The trouble with CAT is that it is so very complicated. Is it essentially a model sustainable enterprise? A museum of eco-gadgetry? A showcase for natural landscaping? An against odds tourist attraction? An idealistic working community? A hands-on training centre? A high-octane eco-salon? A sunlit destination for pilgrimage?
>
> (Harper 2014: 1)

It was all of these things and more. At different times and by different people these multiple strands were emphasised – sometimes clashing and at other times working in harmony.

What the Locals Thought

Neither the local community of the Dyfi Valley nor CAT itself are homogenous communities. The local community is divided between Welsh speakers and English speakers. It is divided along class lines, with landowners such as John Beaumont, farmworkers and the children of redundant slate workers. There are English downsizers and retirees who had moved to find a more rural and more peaceful lifestyle, and Welsh sheep farmers struggling to make ends meet. The slate industry, one of the largest employers in the region, was in decline. Llwyngwern Quarry, where CAT is located, was founded probably around 1835, and by 1900, it was employing 144 men. It ceased production in 1941, although there was an unsuccessful attempt to restart the quarrying in 1950 (Parry 2011). There had been a major exodus from mid-Wales not only because of the collapse of slate quarrying, but also because demand for agricultural labour fell with increasing mechanisation. There were also increasing employment opportunities in places like the Welsh coalfields further south, which paid more than agricultural wages (see Thomas 1995: 411–12). The demise of the quarrying and the nature of the slate industry that devastated the environment had left a rather desolate overgrow landscape, more reminiscent of an industrial wasteland than a rural idyll. Much has been made in recent years that the activity at the Quarry has actually reclaimed this industrial wasteland and increased biodiversity.

> The fact that with every development that's been done on the site, the biodiversity has been found to actually increase. That has always felt to me incredibly important.
>
> (Sally, interview May 2021)

The early community at CAT at first sight seemed to be more homogenous than the local community as the majority of workers and volunteers were, with a few exceptions, well-educated English middle class. Yet the CAT community was far more heterogeneous than it appears.

> I don't know if there was a single thing that linked everybody at CAT.
>
> (Pete Raine, Oral History)

This diversity at CAT tended to be less determined by demographic differences than a variety of distinct approaches to the perceived environmental damage caused by human activity. If we accept that community is essentially a relational phenomenon, then not only are interactions within communities complex and negotiated, but also perceived differences between communities are even more so. Consequently, the response to the presence of CAT was neither uniform nor static. The local population was very ambivalent about this project established its midst.

Money was short and dress sense and lifestyle sometimes appeared to be quite different to many of the locals. To a certain extent, it was inevitable that the people at CAT were identified as the 'hippies on the hill'. Liz Todd observed that when her two children went to school, they were sometimes teased for being hippies. Liz Todd also recalled that when she got selected for a position in the local library in Machynlleth, the local applicants were quite upset, and a petition was started indicating that, as an outsider, she should not have got the job (Liz Todd, Oral History).

Audrey Beaumont, the wife of the landowner, observed:

> No one wanted that place to even start here in this valley, Ceinwys, Corris. Betty the Mill, who is long dead now took a petition through Ceinwys, Corris, everywhere [in the valley] against them ... They did not understand it. They did not know what it was all about ... To them you were all a lot of hippies.
>
> (Audrey Beaumont, Oral History)

Audrey Beaumont's observations indicate that CAT, and the early environmental movement more generally, had to overcome the association with the counterculture. While it did have a certain hippie appearance, most volunteers and workers did not identify with the 'turn on, tune in, drop out' ethos of the counterculture. However, because of this misperception, CAT was regarded not only as irrelevant but also as a possible threat to the culture of the Dyfi Valley.

Although no one knows who originated the title, some locals began to refer to the Centre as 'The Shit and Wind'. This not only refers to the experiments on alternative ways of dealing with human sewage and wind power, but also had connotations that the ideas espoused at CAT were 'bullshit' and vacuous. Graham Preston, who was living locally applied for a job at CAT and who was one of the very few of the early workers at CAT to socialise with the local people of Machynlleth, observes that the members of CAT and locals were 'two totally different groups of people with totally different ideas' – and this created very much an 'us' and 'them' situation. Recalling his first week working at CAT, he observed:

> The people, their language was different to anything that I had heard before. Just looking at the people was enough – their dress, the way people were conducting themselves, the things that they were talking about was so different.

> (Graham Preston, Oral History)

In addition, Wales has a very strong cultural tradition, to which many of the early pioneers were not particularly sensitive. A sense of 'us and them' was compounded by the historical relationship between England and Wales. Don Bennett who, though originally from Glamorgan in South Wales, moved to the Machynlleth area at about the same time as CAT was established suggested that English people had been coming across the border into mid-Wales for holidays since the end of World War II. He observed that 'the first holiday cottage people were not very nice [they were] very middle class … There is a deep-rooted suspicion of a certain class of people down here prior to CAT coming down.'

For CAT, there was a sense that saving the planet was an absolute imperative and had to take precedence over the perceived parochialism of Welsh culture. This attitude did not endear the embryonic community with some of the local Welsh. Meri Wells, a local ceramicist with an international reputation, who worked briefly at the Quarry Shop, a vegetarian café and wholefood store that CAT opened in 1979 in the centre of Machynlleth, recalls that she was asked not to speak in Welsh with a local schoolgirl who was helping her in the shop. She resigned at that point. In her oral interview, she observed that this insensitivity to the bilingual culture entailed that 'it was saving the planet but squashing a culture' (Meri Wells, Oral History). This idea that these predominantly English incomers were at best indifferent, if not actually hostile to Welsh culture, also was consistent with the historical relationship between England and Wales. Meri Wells suggests that, until recently, many people at CAT were unaware of the distinctive Welsh culture. However, she and others also acknowledge that CAT has made a significant contribution in raising the consciousness of many locals about environmental issues.

Another misperception about CAT, not only by the local community, but also from the wider public, is that CAT advocated vegetarianism. This perception was perhaps exacerbated by the fact that the Quarry Shop and Café sold only vegetarian food. Much of the land, due to the terrain, around the Quarry is dedicated to sheep farming, and Welsh lamb has also got a special place in Welsh cuisine, culture and economy. However, while all communal meals were vegetarian, by no means all the people at CAT were vegetarian, and while it has tended to suggest that it is better for the environment and for individual health to reduce meat consumption, CAT has never promoted an exclusively vegetarian diet.

Some local people were also a little perplexed about the nature of the project. Many local people remembered the days when not only remote farms but also the town of Machynlleth itself was not connected to mains electricity. There was a local small hydro scheme that generated enough for lighting in the town, which was later supplemented by diesel generators when the demand grew with more people having electrical domestic appliances. In 1961, the Merseyside and North Wales Electricity Board (MANWEB) connected mid-Wales to the national grid. Edward Jones, a local farmer, who both worked at CAT and was one of the locals who was more sympathetic to the project, observed that electricity supplied by the national grid was cheaper, easier, more convenient and available 24 hours a day. At the time, it seemed to the locals that generating electricity through renewables and not being connected to the electric mains was anachronistic and inconvenient.

However, not all locals were hostile or perplexed by CAT. There was also a great deal of support from the very early days.

> There were some delightful local people who took an interest in what we were doing. A number of local people locally used to come and help. They came with tractors to move things and helped dig holes … [However] some people did feel a certain anxiety that these newcomers were going to water down the traditional culture.
>
> (Bob Todd, Oral History)

Edward Jones was intrigued by CAT and began to do contract work at the Centre. Jones was not only fascinated by the engineering side of AT, but also he comments on how people came from all over the world 'and it was very interesting to chat to some of them'.

> People were prepared to work just for their food and lodging. That was very similar to my childhood days on farms where people would help each other, very often just for a meal. So there was not a lot of money exchanged. I enjoyed that. It was a nice way of working and meeting people.
>
> (Edward Jones, Oral History)

Delyth Rees, a local resident of Machynlleth who became a trustee of the charity, recalled that both his grandfather and great grandfather had started cooperatives and grew their own food and recalls that both his mother and grandmother recycled clothing (Delyth Rees, Oral History). Consequently, the values of Welsh culture and those of CAT are in fact far more consistent than it first appears. Andy Rowland, who came to run the bookshop in 1984 and later became the site manager and who was one of the few people at CAT in the early 1980s who successfully became fluent in Welsh, observes that he perceived a commonality between CAT culture and 'Welsh socialist pacifist, land-based traditions' (Andy Rowland, Oral History).

There was perhaps little awareness and consideration in pioneering days about the impact of the Quarry, which initially attracted mostly English outsiders, of the local Welsh culture. The focus was mostly on making the Quarry habitable, the larger picture of the looming environmental disaster and developing the visitor centre. Consequently, there was little time for the immediate context and Welsh culture. However, there were a number of significant arenas that have contributed to CAT becoming more integrated with the local community. Perhaps the two most significant, and in some ways interconnected, ways that CAT became more integrated into the local community was when the children of CAT workers went to local schools and when more people from CAT started to learn Welsh. Another way that CAT became more involved with the local community was when a number of CAT staff and Quarry kids started to participate in the Machynlleth pantomime, including directing, acting and stage management. Pete Raine, the second Director of CAT, who directed three of the local pantomimes, regarded it as 'one of the major ways in which the Quarry integrated into the town' (Pete Raine, Oral History). Common ground was also found when the UK Government was seeking sites to store nuclear waste in the early 1980s. Local farmers and other campaigners including some from CAT managed to prevent surveyors from the Institute of Geological Sciences from carrying out their survey to find a suitable site for nuclear waste in mid-Wales.

Pete Raine also indicated that the Quarry Shop and Café in the centre of Machynlleth was also significant in helping CAT integrate into the local community. However, some of the locals were rather wary because it was a vegetarian café and initially seemed rather alien.

> Only a certain number of local people came in [to the Quarry Shop] …
> I would sit with and talk to local people in the pub and they would say 'I would like to come in, but I don't dare. It is too different'. People said to me 'I got to the door, but I just chickened out.
>
> (Nigel Dudley, Oral History)

A local doctor indicated to some of the older locals that bran is an aid to digestion, and the Quarry Shop was the only place in Machynlleth that sold bran, and consequently locals started to come in. However, Richard St George suggests that many of these older Welsh customers thought that the dried herbs that were sold from jars behind the counter was marijuana and 'people used to come in with their eyes closed, get the bran and dash out'. Nonetheless, 'The

Quarry Shop did help make contact with local people' (Richard St George cited in CAT 1995: 21). The Quarry Shop and Café did become popular amongst some locals. Many of the girls from the local school started to come to the Café for their lunch. Although, as Nigel Dudley observes, the boys from the school tended to get chips from the local fish and chip shop and hang outside the Quarry Shop and Café. Eventually, due to lack of space, as both the cafe and shop became more popular, the shop moved to separate premises. Jonathon Gross who ran the Quarry Shop described the shop as a sort of embassy for CAT on the High Street.

> The Quarry Shop is a part of CAT, but it is also definitely. A part of the community as well, and it's actually an integral part of the way that the High Street works.
>
> (Jonathan Gross, Oral History)

After the closure of both the shop and the café in 2012 when CAT was forced to restructure following some financial difficulties, Gross observed that many local people seemed genuinely 'bereft' as the shop and the café were regarded as integral to the culture of Machynlleth High Street. Both the Café and the Shop were reopened by former CAT members of staff as independent businesses.

As both the local and CAT are heterogeneous communities, the interactions between locals and the Quarry were and remain complex, diverse and changing. There was an ambivalent relationship between CAT and the Welsh community. CAT has bought many outsiders to the area, some of whom settled. This is perceived by some as diluting the Welsh culture. In addition, the majority of people working at CAT were outsiders and CAT did not employ many local people except as hourly paid staff in the trading outlets such as the Quarry Shop.

Although the interactions between the local community and CAT were somewhat tenuous, at least in the early days, CAT has had an impact on the area. Pete Raine, who is a qualified accountant, calculated that CAT contributed 5% of the GDP of the Dyfi Valley.

> The area is more eco-friendly than it might well have been had it not been for CAT being here. There are many more fair trade and organic products in the local stores and recycling came to Machynlleth sooner than to other small towns in mid-Wales.
>
> (Meri Wells, Oral History)

In later years, this percolation of green ideas out to the wider community manifests in a more concrete way with the formation of Ecodyfi in 1998 'an independent social enterprise delivering sustainable community regeneration and promoting a growing green economy in the Dyfi Valley' (Ecodyfi 2020). Bro-Dyfi Community Renewables was also started in the area. It describes itself as 'green power and local ownership in action'. Bro-Dyfi is a community-financed renewable energy scheme in which local people invested in a wind turbine. There was so much enthusiasm from the local people that they invested in a second turbine. The power generated is sold to suppliers, makes a small dividend for the investors and also helps finance small local sustainability projects. The Dyfi Valley was also designated by UNESCO as a biosphere reserve in 2009. 'UNESCO biosphere reserves promote solutions reconciling the conservation of biodiversity with its sustainable use. They are learning areas for sustainable development under diverse ecological, social and economic contexts' (UNESCO 2021). Andy who worked at CAT and later for Ecodyfi suggested that probably none of these initiatives would have happened without CAT (Andy, interview August 2021).

While some locals and many in the wider society perceived the mores of CAT as radical and alien, those involved at CAT believe that it is imperative that the practical solutions they advocate are adopted by all if the global environment that sustains all life is to be preserved. Consequently, education and outreach are critical for the environmental movement in general and plays an absolutely essential aspect of CAT's agenda from the very beginning to its current incarnation. However, neither the wider context nor CAT are static. In the late 1980s and early 1990s, environmental discourses began to shift, and CAT responded to this changing context in what CAT called Gearchange.

Gearchange

In the late 1980s, there was a shift in the discourses about environmentalism. In the 1970s and early 1980s, it was generally thought that environmental protection and economic growth were mutually exclusive. The idea of economic growth and quality of life was measured in economic terms and was considered to be the *sine qua non* of modernity. Environmentalism, as exemplified by *The Limits to Growth*, was seen as inherently antithetical to economic growth and therefore retrogressive and detrimental to levels of comfort and convenience created by modernisation. This concept that it was a choice between economic growth and protection of the environment began to shift in the late 1980s following the emergence of what is often referred to as ecological modernisation. Stephen C. Young suggests that ecological modernisation entailed 'reconceptualising the relationship between the environment and the economy' and that 'growth could be maintained within a framework of stronger environmental protection' (Young 2000: 2). There was a dawning realisation by both government and business that there was not necessarily an inverse relationship between environmental protection and economic growth. For example, there was a growing awareness that it was in industry's best interest to develop less polluting and less resource-intensive technologies. This is what might be called the greening of capitalism.

Another important aspect of ecological modernisation was the idea of sustainable development. The most important articulation of this idea was the World Commission on Environment and Development (WCED) report *Our Common Future* of 1987, more frequently called the *Brundtland Report*, after the

Commission's Chair Gro Harlem Brundtland. This report contains the most cited and influential definition of sustainable development:

> Sustainable development is development that meets the needs of the present without compromising the ability of future generations to meet their own needs.
>
> (WCED 1987, Chapter 2, point 1)

It is often said that the authors of this report invented the concept of sustainable development. Dryzek (2005: 147) notes that sustainable development 'is arguably the dominant global discourse of global concern'.

The implications of the development of ecological modernism and the concomitant discourse of sustainable development had two major interrelated consequences for CAT and the environmental movement more generally. First, this is the beginning of environmentalism coming in from the cold and becoming a mainstream concern. Radkau suggests that environmentalism that 'had previously been thought of as a hillbilly phenomenon now became champions of the common good and pioneers of a new modernity' (Radkau 2014: 353). The second was that the link between environmental protection and self-sufficiency became less tenable as governments, business and industry started to realise that environmental issues were of national and international concern. This changing context led to a growing awareness that CAT was in danger of becoming anachronistic, and that changes were required to make it relevant and appealing in the context of environmental modernisation.

In addition, there was a financial imperative to widen CAT's appeal. CAT receives no government funding and received very little in external grants at all, as there was surprisingly no fund-raising department.[21] Income was primarily derived from the commercial and educational aspects of CAT. This included entrance fees to the site from day visitors and fees from school visits and courses. The Quarry Café and Shop, a restaurant and a bookshop on-site, and a thriving mail order business selling books and various green products also contributed to CAT's income.

In the late 1980s, CAT felt that because environmental issues were much more in the public domain, there was a potential to increase visitor numbers. More visitors would not only make a more substantial contribution to CAT's finances but could also raise the awareness of more people. Roger Kelly, who was the Director of CAT at the time, recalled that there was a debate about the target audience for CAT's message:

> One of the big issues, certainly in my mind and I think in a lot of people's minds, whether you are trying to affect a large number of people, but at a fairly shallow level or a much smaller number of people at a much deeper level. And we opted for the first of those. So, we thought it was most important to get as many people as possible on the sort of first rungs of the ladder.
>
> (Roger Kelly, Oral History)

Furthermore, as environmental issues were no longer marginal, it was agreed that the visitor centre required a major overhaul. Tim Kirby, the site engineer at the time, observed:

> That change had to happen. The world was changing. Even Maggie Thatcher was making green speeches by then. The old spit and sawdust Quarry would not have kept on attracting visitors. Also there needs to be a gearchange in thinking, because pre-Gearchange the idea of wind to generate electricity was new to most people. Now there's been an explosion of information and a lot of the messages that Quarry was trying to get over were suddenly out.
>
> (cited in CAT 1995: 38)

Similarly, Clive Newman, another of the engineers, observed that:

> The environmental movement was changing, and we had to respond. It was quite nice doing little amateurish things on the hill, but we had to get a bit smarter, a bit tidier and better. Gearchange was a response to that.
>
> (Clive Newman, Oral History)

One of the problems that CAT faced, as well as being located in a fairly remote part of rural Wales, was access to the site itself. There is a fairly substantial-sized car park just off the main road between Machynlleth and Dolgellau, but visitors then had to walk up a steep track to get to the site. In meetings, it was suggested that a water-balanced cliff railway could be a possible way of getting visitors the 30-metre ascent to the site. This would be an attraction in itself as well as facilitating access.

> The whole idea was not only to get people up to the top in a relatively painless way, but also to demonstrate renewable technologies as well.
>
> (Cindy Harris, Oral History)

Not only would a water-balanced cliff railway be consistent with the technological ethos of CAT, but it would also have links with the slate mining industry, that used a similar technology to get slates from the Quarry down to the main road. The water-balanced design has two carriages on parallel tracks up the slate tip. The one at the top fills with water so that its weight is more than that of the passengers in the lower carriage. This extra weight allows the descending carriage to pull the other carriage up the slope. When the carriage gets to the bottom of the slope, it empties the water, passengers board and the process is repeated. The water-balance system required building a reservoir near the top station. This was beautifully landscaped and has really added to the overall look and ecological diversity of the site.

The challenge was to raise the money to finance this ambitious scheme as CAT had never done anything on this scale before. It was also a reversal of the previous model of financing, which was primarily 'we have the funds, what are our spending priorities?'. In contrast, Gearchange was a proposal that required CAT to raise the funds for a specific project. Ironically, raising the finances took advantage of a scheme introduced by the then Prime Minister Margaret Thatcher, which gave tax breaks for purchasing shares in start-up companies. CAT formed a Public Limited Company (PLC) that would run parallel to the charity and launched an ethical share issue in 1990. The share issue was not anticipated to pay dividends as such but was reliant on sufficient people who thought CAT to be a worthy cause. The goal was to raise a million pounds within a year, and CAT actually achieved this in about 18 months. Roger Kelly, the Director at the time, wrote in the introduction to the prospectus for the share launch:

> This is an invitation to participate financially in the expansion of the Centre for Alternative Technology plc, not as an investment for personal profit, but out of commitment to an environmental initiative.

This is not a usual investment, but an invitation to invest in the future of the planet, rather than any personal financial gain. However, the prospectus notes that while visitor numbers had remained fairly static at about 50,000 per annum in the eight years up to 1988, CAT had seen an increase in visitor numbers in 1989 and projected that with the proposed cliff railway and improvements to the visitor centre that this could reach 150,000 visitors in 1994. Roger Kelly suggested that 'we saw the day visitor as being what could both get the message out to many more people and bring in the money to keep the place afloat.'

Having successfully raised a million pounds, work on the cliff railway began. This included building the timber framed top and bottom stations, making a lake at the top to supply the water necessary for the operation of the water-balanced railway. This work was completed in 1991. The stations for the cliff railway were built in as environmentally friendly way as possible (see Chapter 4).

Initially, visitor numbers did increase to perhaps 90,000 at its height, but soon plateaued and started to decline. It remains unclear why visitor numbers declined, and there has been a great deal of speculation as to the cause for the fall in footfall. Towards the end of his time as Director, Roger Kelly speculated that the CAT as a visitor centre had run its course. Both the discourse about the environment and the wider socio-political context was very different in the late 1990s and early 2020s. In addition, the technology itself had also moved on. By this period, environmental issues had moved very much from the margins to the mainstream. Many of the ideas espoused by CAT were no longer 'alternative' as such. Wind farms were beginning to appear in various parts of the UK. Although there was some backlash, with mostly people living near complaining about noise, spoiling the landscape or wrongly assuming that wind turbines posed a threat to birdlife. Photovoltaics were becoming better and cheaper could even be effective in the UK. The technology was no longer the DIY style of early CAT. Renewables were no longer strange and exotic, and the focus was less on small scale and the idea of self-sufficiency of John Seymour. In just a couple of decades, wind turbines had advanced from the few kilowatt turbines to the massive machines that produce several Megawatts of power. We have come a long way from the Cretan Windmill and the 12 kilowatts of power that used to be what powered the entire site.

Part of the Gearchange project was the decision to connect part of the site to the national grid. This was in part because of practical reasons. The decision to connect to the grid was primarily advocated by the engineers. It was difficult for the non-engineers to contradict the technical argument for connecting to the grid. The debate was finally decided, when the Railway Inspector decreed that the cliff railway needed a connection to the grid for health and safety reasons. However, in typical CAT fashion, a compromise was agreed that other than the top station area the rest of the site would remain off-grid. There used to be a line of green turf – colloquially known as 'the green line' – that demarcated the point where the site became reliant on its own systems. However, by the early 2000s, with the rise in visitor numbers, increasing staff and the necessity for more computers, it was becoming increasingly challenging for CAT to be self-sufficient in power.

> As demand increased, we had to run the back-up diesel more and more, and people increasingly asked, if we are producing electricity inefficiently from fossil fuels on-site, why don't we get it more efficiently (and cheaper) from the grid.
>
> (Peter Harper, personal correspondence February 2022)

By the early 2000s, there was also an opportunity for CAT and other producers to sell renewable energy back to the grid.

> We just saw the grid as something that we could use to put in as much renewables as we could and take out little bits when we needed to.
>
> (Cindy Harris, Oral History)

However, after connecting to the grid, 'there was a lot less awareness of the necessity to limit energy use' (Cindy Harris, Oral History). Some saw this connection as a compromise too far, that CAT was no longer walking the talk. However, the connection to the grid in many ways symbolises the move from more inward-looking bonding social capital to the more outward-looking bridging capital. Bonding social capital was significant in the formation of the community aspect of CAT, which I will discuss in the following two chapters on community and consensus decision making. Bridging capital is more significant in the outreach aspects of CAT, which I will cover in the chapters on outreach and education, and the Zero Carbon Britain project.

Notes

1   However, John Muir is often referred to as a 'preservationist' rather than a 'conservationist'. This distinction arose from the debate between Muir and Gifford Pinchot. Muir was intent on preserving wilderness areas, whereas Pinchot had a more managerial approach to conservation. This all came to a head in the controversy over building a dam in the Hetch Hetchy Valley, which is located in Yosemite National Park, to provide water for San Francisco. Muir vehemently opposed this, but it was supported by Pinchot and the dam was built in 1913.

2   There are of course many other publications that are important to the early environmental such as Murray Bookchin *Our Synthetic Environment* (1962), Paul R. Ehlrich *The Population Bomb* (1968), Barry Commoner *The Closing Circle* (1971) and Ivan Illich *Tools for Conviviality* (1973) – but none had the reach or influence that *Silent Spring, The Limits to Growth* or *Small is Beautiful* had.

3   A reference to the New Alchemy Institute in Cape Cod – one of the places that possibly inspired the founding of CAT.

4   This is the term used by Murray Bookchin.

5   ITDG changed its name to Practical Action in 2005.

6   Later renamed as the Energy and Environment Research Unit.

7   Peter Harper eventually worked and lived at CAT; however, Godfrey Boyle never did, but continued working at the Open University.

8   Now called The Green Centre www.thegreencenter.net/

9   The formal name of the charity was changed to the Centre for Alternative Technology Charity Limited in June 1990.

10   Held in the CAT Archives at the National Library of Wales Box 1/12.

11   This building became the restaurant.

12   CAT Diary (CAT Archives Box 2/7).

13   CAT Diary (CAT Archives Box 2/7).

14   Tony Williams stayed for only a few months and was replaced as site manger by Mark and Mary Mathews who had arrived in March 1974.

15   This was probably Steve Boulter. In *Undercurrents* Volume 7, it was reported that Steve Boulter left as there was some fundamental disagreement on the direction of CAT.

16   21 February 1974 – CAT diary (CAT Archives Box 2/7).

17   7 March 1974 – CAT diary (CAT Archives Box 2/7).

18   Todd was not the only person with an impressive academic background and significant skill set. Jeremy Light, who came in 1976 to organise the biological aspects, has a PhD in Biology and had been part of the Antarctica Expedition. Tim Kirby who came to work as the site engineer in 1982 has a Degree in Engineering Science from Oxford University and considerable engineering and managerial experience – the list could go on.

19   This building was erected on-site that became known as Tea Chest and it became the hub of the site community.

20   CAT Archives Diary 18 October 1974.

21   CAT still receives no government funds. However, in recent years, CAT has been more effective raising grant money from external bodies and funding agencies, and has a fund-raising department.

References

Alpanda, S. & Peralta-Alva, A. (2010). Oil Crisis, Energy-Saving Technological Change and the Stock Market Crash 1973–74. *Review of Economic Dynamics* 13, 824–42. doi: 10.1016/j.red.2010.04.003

Bardi, U. (2009). Peak Oil: The Four Stages of a New Idea. *Energy* 34, 323–6. doi: 10.1016/j.energy.2008.08.015

BBC Sounds. (2013). *The Reunion: The Centre for Alternative Technology*. Available at: www.bbc.co.uk/sounds/play/b01s393k (Accessed 2 December 2020).

Boulter, S. (1974). A.T Goes Boom. *Undercurrents* 7, 4. Available at: https://issuu.com/undercurrents1972/docs/uc07_jan20a (Accessed 10 December 2020).

Boyle, G. (1974). The Princely Pursuit of A.T. *Undercurrents* 8, 1. Available at: https://issuu.com/undercurrents1972/docs/uc08_jan20a (Accessed 21 January 2021).

Boyle, G. & Harper, P. (Eds.) (1976). *Radical Technology*. London: Wildwood House.

Carson, R. (2000). *Silent Spring*. London: Penguin Classics.

Centre for Alternative Technology (CAT). (1995). *Crazy Idealists: The CAT Story*. Machynlleth: CAT Publications.

Charity Commission for England and Wales. (2008). *Campaigning and Political Activity Guidance for Charities*. Available at: www.gov.uk/government/publications/speaking-out-guidance-on-campaigning- and-political-activity-by-charities-cc9 (Accessed 9 September 2021).

Clapp, B.W. (1994). *An Environmental History of Britain Since the Industrial Revolution*. London: Longman.

Dickson, D. (1974). *Alternative Technology and the Politics of Technical Change*. London: Fontana.

Doherty, B. (2002). *Ideas and Actions in the Green Movement*. London: Routledge.

Dryzek, J.S. (2005). *The Politics of the Earth: Environmental Discourses*. Oxford: Oxford University Press.

Dunn, P.D. (1978). *Appropriate Technology: Technology with a Human Face*. Basingstoke: Macmillan.

Ecodyfi. (2020). Vision and Aims. Available at: www.ecodyfi.wales/vision-and-aims (Accessed 15 January 2021).

Eldred, K. (1989). Promise Rediscovered: New Alchemy's First Twenty Years. *New Alchemy Quarterly* 37, 4–18. Available at: www.thegreencenter.net (Accessed 2 January 2020).

Halfacree, K. (2006). From Dropping Out to Leading On? British Counter-Cultural Back-to-the-Land in a Changing Rurality. *Progress in Human Geography* 30(3), 309–36. doi: 10.1191/0309132506ph609oa

Harper, P. (2014). Songs of Experience: Four Decades of Successes and Failures at the Centre for Alternative Technology. Originally published in *Whole Earth* in 2002 and updated in 2014. (Unpublished paper supplied by the author).

Harper, P. (2015). The Centre for Alternative Technology: An Extreme NGO. Paper presented at the *Oxford Centre for Humanities* July 2015. (Unpublished and kindly supplied by the author).

Harper, P. (2016). Alternative Technology and Social Organisation in an Institutional Setting. *Science as Culture* 25(3), 415–31. doi: 10.1080/09505431.1164406

Harper, P. & Erikson, B. (1972). Alternative Technology: A Guide to Sources and Contacts. *Undercurrents* 3 (Autumn/Winter), 17–26. Available at: https://issuu.com/undercurrents1972/docs/uc03_jan19b (Accessed 4 December 2020).

Harrabin, R. (2014). Turning a Slate Quarry Green: 40 Years of Centre for Alternative Technology. *The Guardian.* (1 August) Available at: www.theguardian.com/environm ent/2014/aug/01/turning-a-slate-quarry-green-40-years-of-centre-for-alternat ive-technology (Accessed 19 August 2021).

Heidegger, M. (2013). *The Question Concerning Technology.* New York: Harper Perennial Modern Thought.

Jacob, J. (1997). *New Pioneers: The Back-to-the-Land Movement and the Search for a Sustainable Future.* University Park: The Pennsylvania State University Press.

Killingsworth, M.J. & Palmer, J.S. (1996). Millennial Ecology: The Apocalyptic Narrative from *Silent Spring* to *Global Warming.* In C.G. Herndl & S.C. Brown (Eds.), *Green Culture: Environmental Rhetoric in Contemporary America.* Madison: University of Wisconsin Press, 21–45.

Kirk, A.G. (2007). *Counterculture Green: The Whole Earth Catalog and American Environmentalism.* Lawrence: University of Kansas Press.

Macnaghtan, P. & Urry, J. (1998). *Contested Natures.* London: Sage.

Maher, N. (2004). Shooting the Moon. *Environmental History* 9(3), 526–31. doi: 10.2307/ 3985771

Meadows, D.H., Meadows, D.L., Randers, J. & Behrens III, W.W. (1972). *The Limits to Growth.* New York: Universe Books.

Mirzoeff, N. (2015). *How to See the World.* London: Pelican.

Morgan-Grenville, G. (1973). *The Society for Environmental Improvement. The National Centre for Alternative Technology.* Unpublished paper. Available in the CAT Archives at the National Library of Wales (Box 1/12).

Morgan-Grenville, G. (1974). Interview: The National Centre in *Undercurrents* 8, (October/November), 12–14. Available at: https://issuu.com/undercurrents1972/ docs/uc08_jan20a (Accessed 4 December 2020).

Morgan-Grenville, G. (2001). *Breaking Free.* Bridport: Milton Mill Publishing.

Muir, J. (2019). Our National Parks in John Muir (Ed.), *The Wilderness Essays.* Praha: Czech Republic: Madison and Adams Press, 425–918.

National Center for Appropriate Technology (NCAT). (2021). *About NCAT.* Available at: www.ncat.org/about-us (Accessed 9 September 2021).

Parry, C. (2011). Llwyngwern Slate Quarry. Available at: https://coflein.gov.uk/en/site/ 407582/details/llwyngwern-slate-quarry (Accessed 10 February 2021).

Pursell, C. (2009). Sim Van der Ryn and the Architecture of the Appropriate Technology Movement. *Australasian Journal of American Studies* 28(2), 17–30.

Radford, T. (2011). *Silent Spring* by Rachel Carson Review. *The Guardian* (30 September). Available at: www.theguardian.com/science/2011/sep/30/silent-spring-rachel- carson-review (Accessed 25 August 2021).

Radkau, J. (2014). *The Age of Ecology.* Cambridge: Polity.

Rybczynski, W. (1991). *Paper Heroes: Appropriate Technology: Panacea or Pipe Dream.* London: Penguin.

Schumacher, E.F. (2011). *Small is Beautiful: A Study of Economics as if People Mattered.* London: Vintage Books.

Shamsuddha, M. (2017). *An Analysis of James Lovelock's Gaia: A New Look on Life.* London: Macat.

Slack, J.D. & Wise, M.J. (2015). *Culture and Technology: A Primer.* New York: Peter Lang.

Smith, A. (2005). The Alternative Technology Movement: An Analysis of its Framing and Negotiation of Technology Development. *Human Ecology Review* 12(2), 106–19.

The Diggers. (1968). Trip Without a Ticket. *The Digger Papers*, 3. Available at: www.digg
ers.org/digger_papers.htm (Accessed 19 August 2021).

The Schumacher Institute. (2021). *Schumacher Circle*. Available at: www.schumacherin
stitute.org.uk/about-us/schumacher-circle (Accessed 1 September 2021).

Thomas, I.C. (1995). The Beginnings of an Economic Development Policy in Mid-
Wales: The Mid-Wales Industrial Development Association 1957–1974. *Welsh
History Review* 17(3), 411–51.

Undercurrents. (1972). Science with a Human Face. *Undercurrents* 01(February),
2. Available at: https://issuu.com/undercurrents1972/docs/uc01_dec19b (Accessed
13 September 2021).

UNESCO. (2021). *Biosphere Reserves*. Available at: https://en.unesco.org/biosphere
(Accessed September 2021).

Urry, J. (2011). *Climate Change and Society*. Cambridge: Polity Press.

Ward, B. (1966). *Spaceship Earth*. New York: Columbia University Press.

Wiener, A. (2018). The Complicated Legacy of Stewart Brand's 'Whole Earth Catalog'.
New Yorker (16 November) Available at: www.newyorker.com/news/letter-from-sili
con-valley/the-complicated-legacy-of-stewart-brands-whole-earth-catalog?utm_
source=NYR_REG_GATE (Accessed 23 August 2021).

Williams, R. (1988). *Keywords: A Vocabulary of Culture and Society*. London: Fontana.

World Commission on Environment and Development (WCED). (1987). *Our Common
Future*. Available at: https://sustainabledevelopment.un.org/content/documents/
5987our-common-future.pdf (Accessed 21 September 2021).

Young, S.C. (Ed.) (2000). *The Emergence of Ecological Modernisation: Integrating the
Environment with the Economy*. Oxon: Routledge.

Zelko, F. (2013). *Make it A Green Peace: The Rise of Countercultural Environmentalism*.
Oxford: Oxford University Press.

2 From Community to Network

Introduction

In this chapter, I will discuss the idea of community, and whether this can be considered as helping or hindering the Centre for Alternative Technology (CAT)'s agenda to promote a sustainable lifestyle. On the one hand, being able to demonstrate actually living a low environmental impact lifestyle contributed to CAT's credibility – here was a group of people not only advocating sustainability, but also actual attempting to live the life. In other words, the community itself can be perceived as part of CAT's rhetorical strategy that supplemented the more discursive aspects of its argument. On the other hand, the community also had the potential to detract from the persuasiveness of CAT's message. The emphasis on community cohesion and bonding capital, which I discuss below, could attenuate the imperative for developing bridging capital. It was also sometimes suggested that a sustainable lifestyle was only apposite for a small group of people living in a remote rural location, and that these ideas could not be readily transferred to an urban context. In other words, while the communal aspects of the CAT project suggested a viable alternative, this alternative could be perceived as not being achievable for most people.

The Site Community

To set the scene, in the very early pioneering days, almost everybody lived on-site in caravans. This was a matter of practicality as it was more convenient to be on-site while work was done clearing the site and restoring the slate workers' cottages. The very early pioneers received little more than pocket money and unless individuals had savings or another source of income, they could not afford to live elsewhere even though at the time house prices and rents were relatively cheap compared to most other parts of the UK. Living on-site also fostered a sense of community.

In the first year, the population of CAT was very fluid. Prior to being appointed as Director, Roderick James, who was living in the vicinity, went up to the Quarry on several occasions to mentor the volunteers in various building

DOI: 10.4324/9781003207702-3

tasks. Most of these volunteers were very enthusiastic but lacked much in the way of practical skills.

> There was a continuous stream of people coming through. I would show a group how to make a window frame and then ten days later those people were all gone and there would be another lot of people there.
>
> (Roderick James, Oral History)

However, once James had been appointed the Director and several key figures had responded to the advertisement for anyone who had time to spare, a fairly stable core of people was established. However, there was always a shifting population around this stable core.

After the first pioneering year, it was apparent that the on-site community was not for everyone. James never lived on-site, as he had a property nearby. Some found properties in the vicinity preferring to live off-site. As CAT expanded, there was not sufficient accommodation for everybody, and some people who came to work at CAT in subsequent years had no desire to live on-site. Consequently, there was a distinction between those living at the Quarry, generally referred to as the site community, and those who did not. When I was there in the mid-1980s, the site community was a strong and vibrant aspect of CAT. There were approximately 30 permanent members of staff, just over half of whom lived on-site. In addition to the permanent site members, there was also a constant flow of volunteers (often simply referred to as vols). There were short-term volunteers who only came for a week, and long-term volunteers who came for six months and became temporary members of the site community. Eventually, the site community was wound up in 2010. In his oral interview, John Urry, who worked at CAT for over 30 years, and his family were the last people living on-site, surmised that perhaps the importance of the site community had now passed.

In the very beginning, the idea of the community was a significant aspect of the CAT project. It is clear in the position paper written by Morgan-Grenville, just after the acquisition of the site, that the community was an important aspect of his vision. He planned:

> The establishment of an autonomous community where the results of the research at Llwyngwern can be applied in a practical form. The purpose of this will be, not only to give those who wish to adapt their life-styles to such a form the opportunity to do so, but also to provide a demonstration of the relevance of such life-styles to any long-term reduction of environmental problems.
>
> (Morgan-Grenville 1973)

In other words, CAT was intended to not only be a place to test and develop the then quite rudimentary alternative technology (AT) but would also be an experiment in sustainable living. However, the role of the site community was never fully resolved. Was it simply to provide accommodation or, as Morgan-Grenville envisaged, did the site community have a more integral role to play in the project? The site community did give CAT a certain credibility as it was a demonstration of sustainable living. The site community was a demonstration of what is possible and that CAT practiced what it preached. On the other hand, the community aspect could also be alienating – not everybody wants to or is able to live in a small community.

At one end of the site, there is a terrace of three cottages, a small single storey building that was originally the Quarry manager's office and a storage shed, all built of slate. These were eventually restored and provided living spaces for the community. Three large caravans were at the other end of the site.[1] Right in the middle of the site is a communal building called Tea Chest, which was donated to the community. This kit house was originally displayed at The Ideal Homes Exhibition 1974. Nobody really knows why this building is called Tea Chest, although there is some speculation that it was either because it was more or less flat-packed for transport in wooden crates that resembled tea chests or because it vaguely looks like a tea chest as it is clad in wood. Tea Chest had a shared kitchen and provided dormitory accommodation for the short-term volunteers. Tea Chest was not only physically at the centre of the site, it was also the heart of the community.

The significance of the community is dependent upon whom you talk to. For some, the site community was peripheral at best, for others it never fulfilled its purpose as part of the demonstration of AT. However, for others, the site community was integral to CAT. Tim, a member of the site community from 1982 to 1986, suggested:

> The teaching and the outreach was very important, but equally important was the community that was looking to live with the ideals that CAT preached ... The community was very much part of the experience of being there.
>
> (Tim, interview February 2021)

The questions that I want to address in this chapter are: what is community and is a community a requirement for addressing environmental issues? The answer to the second question is clearly contingent on the answer to the first. As Raymond Williams (1988: 75–6) identifies, the concept of community has a range of different meanings, from the geographical that signifies a group of people who inhabit a particular region, to a sense of common identity. While community has a wide range of often contested meanings, I agree with Gerard Delaney's assessment that what underpins all the very diverse conceptualisations of community is a sense of belonging and sharing (Delaney 2018: 5).

Despite some divisions, I argue that there was a wider CAT community that included both on-site and off-site members of staff, as there was a strong sense of belonging and sharing that extended beyond the site community. Belonging and sharing are not absolute, and there are degrees of belonging and sharing.

Listening to the oral histories, the sense of sharing is a very common theme. It was not so much the sharing of tangible objects, but a sense of shared responsibility and vision, that was fostered not only through an environmental sensibility but also through consensus decision making and rotas for many things such as cooking and cleaning toilets, which I will discuss more fully in Chapter 3.

What is CAT: Commune, Cohousing, Ecovillage or Intentional Community?

The CAT community has elements of an ecovillage, an intentional community and cohousing, though formally was none of these. There was also a wrongly held perception that CAT was a sort of commune. While there was some communal sharing, CAT is not and never has been a commune. However, David Pepper includes CAT as a case study in his 1991 publication *Communes and the Green Vision*. Although Pepper lists ten defining features of a commune, he acknowledges that some of the communities discussed in his book 'do not fall into this definition of communes' (Pepper 1991: 5), so one wonders why Pepper uses the term 'commune' at all. Other than the first criterion that membership is voluntary, all of Pepper's characteristics of communes are far more equivocal or do not apply to CAT. For example, Pepper suggests that communes 'are relatively withdrawn from the wider society' (Pepper 1991: 5). CAT is located in a fairly remote rural location, but it would be an oversimplification to suggest that it has withdrawn from wider society. In fact, the opposite is true as arguably the most significant raison d'être of CAT is to reach out to the wider society. One of the aims of CAT stated in the Memorandum of Association (2010) is to educate the wider public about sustainability, and this necessarily precludes the idea of establishing a sort of hermitic community. Consequently, a more precise definition of a commune is required. Communes do not tend to have the same legal status as cooperatives. Communes are mostly associated with the countercultural experiments of the 1960s, and hence the hippie label that CAT acquired. For example, when Liz Todd recollects that when she got the job in the library, a local magazine questioned whether they really wanted the librarian to be 'the sort of person who eats communal meals' (Liz Todd, Oral History).

Timothy Miller acknowledges that there is no consensus on how to define a commune. Nonetheless, Miller suggests that communes are 'residentially-based groups whose members pool most or all of their assets or income and share a belief system or at least a commitment to important core concepts' (Miller 1999: xxi). While there is a residential aspect to the site community and some sharing of beliefs and resources, there is no pooling of assets or income as such. The assets are formally owned by The Centre for Alternative Technology Charity Limited. Although there was an equal pay system (see Chapter 3), any income was personal and not shared as such. For example, if a partner had an income from elsewhere or a worker had savings, they were not expected to share these with the group. While not a commune as such, there was a strong sense of sharing. This sense of sharing included not only material resources but also responsibilities and ideas, and this extended to off-site as well as on-site participants.

CAT was not a housing cooperative either. The formal definition of a housing cooperative is that accommodation is 'owned and controlled by its members' (Hands 2016: 39). Because CAT is formally a charity, it is open to the public, and a condition of living on-site was that when you were employed by CAT in some capacity, it meant that it was neither owned nor fully controlled by its residents.

The site community, with its combination of private and shared spaces, is perhaps more similar to cohousing than a commune. Cohousing is defined as:

> Intentional communities, created and run by their residents. Each house-hold has a self-contained, private home as well as shared community space. Residents come together to manage their community, share activities, and regularly eat together.
>
> (UK Cohousing 2021)

Virtually all members of the site community indicated that what they valued most about living on-site was the combination of having a private space and also some shared resources.

> Living on-site was a beautiful balance between your own private place and your own private time and being sociable and seeing everybody and sharing things.
>
> (John, interview February 2021)

The site community differed from most cohousing projects in two significant ways. First to live on-site, you had to be employed by CAT, and second, CAT was a public exhibition and educational centre. Many saw the site community as integral to the educational aspect of CAT. In other words, the site community was more than simply providing accommodation, social spaces and shared activities. In addition, as I have indicated, many of those involved at CAT did not live on-site. As CAT staff numbers expanded, the proportion of CAT members living on-site inevitably decreased and the importance of the site community dwindled.

Cohousing inevitably raises the issue of the boundary between public and private spaces, which was also a debate at CAT. Maria Ruiu in her study of cohousing suggested that although there are what she identifies as semi-public spaces that are shared within the community, they are 'theoretically private for outsiders' (Ruiu 2016: 408). However, as CAT was open to the public, the boundaries between public, semi-private and private were not always clear-cut. It is clear that the caravans and cottages were people's homes and therefore private spaces. Although, in the very early days, this was not always recognised.

It was common for people to actually walk in to the cottage and sit down as if they had a right to.

(Liz Todd, Oral History)

However, the space outside houses was rather more equivocal. The ambiguity between the public, semi-private and private was an important dimension in the debate about the purpose and significance of the site community to the wider CAT project. If the site community was integral to the demonstration of AT, then it should be visible to the general public. However, most of the site community valued their privacy and did not want to be on display.

In recent years, the idea of ecovillages has become popular. Jonathon Dawson suggests that ecovillages 'act as centres of research, demonstration and in most cases training' (Dawson 2015: 219). This certainly describes CAT. As the name implies, ecovillages are focused on low-environmental impact housing. CAT does not identify itself as an ecovillage nor is it included as an ecovillage on the Global Ecovillage Network (GEN). Although interestingly, CAT did present itself as 'the village of the future' in the 1980s.

CAT is not a formal housing cooperative, a commune, a cohousing project or an ecovillage, although it did share something in common with all of these. The question that arises is whether CAT constitutes a community, which is obviously a much vaguer and more contested concept. As well as being characterised by a sense of belonging and sharing, a community is often understood in terms of two mutually reinforcing criteria – joining together and setting apart. The sense that one has something in common with some people distinguishes a community from other groups. This sense of commonality is generally determined in spatial terms and/or shared interests. Both these criteria for determining what constitutes a community are problematic. For example, when talking about community in spatial terms, it is unclear what the parameters are – local, regional or national. In addition, there might be more that differentiates my interests, values, beliefs and lifestyle from those of people who live in my immediate neighbourhood, than others who might live on the other side of the world.

Defining a community in terms of shared interest is also problematic. I might well share some interests with another person, but it is unlikely that our interests will totally coincide. For example, I might share my concern for environmentalism with an acquaintance, but we might not share an interest in jazz music. On the other hand, my acquaintances who share a love of jazz might not be particularly interested in environmentalism. Furthermore, sharing is itself a diverse phenomenon and contested concept. Shared interest is a necessary but not sufficient criterion for defining a community. However, it is clear that despite the diverse reasons that individuals have for being at CAT, there is a general concern about both the environment and the organisation itself, and the concept of sharing was a strong aspect of CAT's ethos.

Raymond Williams observes that from the 19th century onwards community was 'the word chosen for experiments in an alternative kind of group-living'

(Williams 1988: 75). However, to distinguish alternative styles of living from other uses of the term, these groups are sometimes identified as intentional communities. In some senses, CAT is an intentional community, which is defined by Barry Shenker (2011: 10) as 'a relatively small group of people who have created a whole way of life for the attainment of a certain set of goals'. In many ways, communes, housing co-operatives and ecovillages might be thought of as types of intentional communities.

Shenker proposes that intentional communities have ten defining characteristics. For example, they are 'founded as a conscious and purposive act' and 'membership is voluntary'. In addition, Shenker suggests that 'sharing is part of the community's ideology', 'the community has collective goals and needs and expects members to work towards their satisfaction' and 'the community's existence has a moral value and purpose which transcends the time-span of individual membership' (Shenker 2011: 10–11). All these characteristics clearly apply to CAT. However, some of the other characteristics of an intentional community identified by Shenker are rather more equivocal. For example, he observes that 'the community is relatively self-contained – most members can potentially live their entire lives in it (or for the period during which they are members)' (Shenker 2011: 11). As indicated, not all participants in the CAT project lived on-site, and while some people stayed on-site for decades, most moved off-site at some point. The unifying factor of membership is not so much where you lived on-site, but that you worked for CAT.

Shenker does acknowledge that the characteristics of intentional communities that he identifies 'can exist not only in varying degrees, but in various ways' (Shenker 2011: 11). Shenker's list and his observation that these characteristics exist to a different extent and in diverse ways indicate a more Wittgensteinian notion of categories. That is, intentional communities as a conceptual category are not so much defined by any single essential feature but by family resemblances. This has two interrelated aspects. First, boundaries between what constitute an intentional community as opposed to any other type of human collectivity, such as society, tribe or subculture, are not clearly defined. Second, there are degrees of conformity. Some groups, for example, the Hutterite community that Shenker discusses, may be paradigmatic exemplars as they clearly exhibit all the characteristics of intentional communities, while other collectivities, such as festival goers, might be much more tenuous. So, while CAT as a whole, including both on- and off-site members, may not be a paradigmatic exemplar, it did exhibit many of the characteristics of intentional communities that Shenker identifies.

Shenker identifies some other significant points in the understanding of intentional communities that can shed light on the nature of CAT. Shenker observes that intentional communities are 'firmly rooted in their society of origin and not in total contrast to it' (Shenker 2011: 240). This in some way can explain CAT's success and longevity. CAT challenges many of the dominant ideas of society; however, it is not so radical that its ideas were totally alien. While there has always been a debate about what depth of green should be promulgated, with the exception of a few of the early individual pioneers, it is fair to suggest that the dominant tone at CAT has predominantly been mid-green. There are, for example, almost no references in the oral archives to writers at the deep green end of the spectrum such as Arne Naess or Murray Bookchin. On the other hand, CAT has always advocated more than light green managerial solutions, such as simply using low energy lighting and doing a bit of recycling, that do not challenge the system. This middle way has been both a strength and a weakness.

> We were sufficiently radical to be heard and to be effective.
>
> (Tim, interview February 2021)

On the other hand, what is radical is also a matter of perspective. In some circles, environmental issues in the UK were associated with the counterculture and the radical left. This perception was not helpful and may have been an obstacle for CAT getting its message across.

Radicalism is also a relationship between a particular community and the wider society. If either the society changes or the community changes, then inevitably that relationship will also change. In the early pioneering days, when self-sufficiency and a hermitic ethos played a greater role at CAT and sustainability was regarded as a fringe concern by the wider society, CAT was perceived as being radical. However, there has been a growing acceptance of many of the ideas that CAT promulgated in the 1970s. Many of the interviewees noted that CAT's ideas are much more mainstream now. Paul Allen, for example, observes that in the 1980s the wider society was beginning to accept many of CAT's ideas.

> We sort of thought, yeah, the green stuff seems to be going in the right direction. We have been on Blue Peter. More and more people are thinking about green stuff and it is beginning to be incorporated in the mainstream.
>
> (Paul Allen, Oral History)

As the perceived gap between CAT and the wider society seemed to narrow, the sense of urgency of the pioneers, informed by the *Limits of Growth, Silent Spring* and *A Blueprint for Survival*, became increasingly attenuated.

There are several common themes that recur in almost all the oral history interviews. Perhaps the most prevalent theme is that no one chooses to work at CAT to get rich. Many of the interviewees talk about the quality of life. In other words, for the vast majority of people, working for CAT is considered as being more than a job. CAT has never been a homogenous group. People at CAT have diverging interests and different understandings of how to address environmental issues. Nonetheless, everyone without exception who has had a permanent post at CAT either implicitly or explicitly expresses that working at CAT is much more than earning a salary and having a nice rural location to live and work, and many suggest that it is more of a mission than a vocation.[2] Furthermore, the consensus decision-making processes, which I will discuss more fully in Chapter 3, fostered a sense of sharing and belonging. These aspects – working at CAT is more than simply a job, CAT has an important and unique mission and the sense of actively participating in the strategic decision making – generate a strong sense of community that cuts across the on-site/off-site distinction and other differences.

Bonding and Bridging Social Capital

These three features – a sense of vocation, having a unique and important mission and a sense of participation – create what Robert Putnam (2000) has called bonding social capital. Robert Putnam draws on the idea of social capital, which he suggests 'refers to connections among individuals – social networks and the norms of reciprocity and trustworthiness that arise from them' (Putnam 2000: 19). Putnam suggests that social capital can be categorised in two forms, which he calls bonding social capital and bridging social capital. Bonding forms of social capital 'are, by choice or necessity, inward looking and tend to reinforce exclusive identities and homogenous groups' (Putnam 2000: 22). On the other hand, bridging capital is 'outward looking and encompass people across diverse social cleavages'. Putnam (2000: 23) continues: 'bonding social capital provides a kind of superglue, whereas bridging social capital provides a sociological WD40'.

> It was the community that was the sort of the gap filling cement that unified a number of disparate people – from individuals who did not want to use nails to university lecturers who were capable of coming up with all sorts of things from sophisticated electronics to energy strategies for whole countries.
>
> (Tim, interview February 2021)

There are two significant points to Tim's observation. First, he indicates the range of responses to environmental issues that can be identified at CAT. Second, he suggests that bonding social capital extends beyond the site community *per se*, yet it is the site community that is integral to creating bonds.

Putnam stresses that 'bonding and bridging are not "either–or" categories into which social networks can be neatly divided, but "more or less" dimensions along which we can compare different forms of social capital' (Putnam 2000: 23). In the early pioneering days of CAT, bonding capital was more significant than bridging capital. This is not to say that CAT was totally isolated. Nonetheless, several of the interviewees in the oral history project suggest that CAT was relatively self-contained and or existed in a sort of bubble.

> It seemed to be a rather strange and isolated place that did not really engage very much with anybody outside.
>
> (Clive Newman, Oral History)

However, as the community and the context evolved, CAT had to develop more bridging capital, not only with the local community, but also with the political establishment, particularly the Welsh Assembly. As Putnam (2000: 23) indicates, 'bridging social capital can generate broader identities and reciprocity'. It is understandable that bonding social capital was important for CAT in the early days to get established and construct a strong sense of identity. However, once this was achieved, it became more important to construct bridges to convince a wider audience that CAT's alternative solutions were both viable and achievable.

One way of thinking about CAT as a community is to consider it as what Putnam calls a vocational community. He speculates that possibly our community ties have moved 'from the front porch to the water cooler' (Putnam 2000: 85). Many places of work construct what Terence Deal and Allan Kennedy (1982) have called corporate cultures. Deal and Kennedy suggest that certain businesses have what they call a strong culture, which is determined by values, heroes, rituals and cultural networks. Values are 'the basic beliefs of an organisation'. Heroes are the people who personify those values and 'provide tangible role models'. Rituals 'provide visible and potent examples of what the company stands for'. The cultural network is the informal '"carrier" of the corporate values and heroic mythology' (Deal and Kennedy 1982: 14–15). I would add mythology to Deal and Kennedy's list of characteristics of a corporate culture. These myths are narratives that relate significant moments in an organisation's history and the role of heroes. These mythic narratives reveal the values of an organisation. Foundational narratives are frequently important for many companies. Changes in the mythic narratives of an organisation can also reveal shifts in focus and values over time. While the development of a strong corporate culture might make for a better workplace and is a useful marketing strategy, it does not necessarily create bonding capital. Putnam points out that there is no evidence to assume 'that the line at the copying machine has replaced the back fence as the locus for social capital in contemporary America' (Putnam 2000: 87).

All the interviewees recorded for the oral history project and many of my own interviewees note that CAT is a very unique place to work. CAT clearly has a very strong corporate culture, with clearly identifiable values, heroes, rituals, cultural networks and mythic narratives. As already noted, while not everybody who has worked at CAT necessarily identified themselves as environmentalists, there was an overall commitment to green values. It is rather contingent on whom you ask as to who is considered to most fully embody those values. Nonetheless, there were a number of names that come up more regularly as 'heroes' in the CAT mythology. In the early days, it was Diana Brass and Gerard Morgan-Grenville who are represented as heroes in the CAT mythology. Brass, in particular, seemed to embody the dedication and pioneering spirit, while Morgan-Grenville represented the vision. Bob Todd is represented as providing the practical skills and scientific credentials that signify the ethos of AT. Perhaps, more controversially, Peter Harper who originally coined the term AT might be considered a hero. While people often, and sometimes vehemently, disagreed with Harper, his commitment to the underlying values of CAT cannot be disputed.

There were numerous ritual events at CAT, which ranged from Monday work meetings to social events that denoted the values of CAT. While there were formal modes of communications, there were also informal networks of communication. One of the most interesting and often cited modes of communication prior to computers was the intercom system that connected most of the cottages, offices and the communal building. The intercoms were housed in small wooden boxes and involved pushing a large red button, which then enabled everyone else near another intercom to hear you. This intercom system was used for a wide variety of reasons: the restaurant urgently requesting an engineer to fix the hot water; one of the on-site families checking if there was enough power to run a washing machine; someone on the phones who had some technical inquiry that they needed help with; or someone struggling with a crossword clue.

Networks of communication not only extended across the site, but socialising was also common amongst on-siters and off-siters.

> We had a respect for each other and a care for each other … we knew more about each other than it in a normal workplace.
>
> (Joan Randle, Oral History)

Similarly, Sabrina Cantor, who worked in various roles at CAT but never lived on-site, notes that:

> Everybody is on the same wavelength … That is the thing about CAT that sense of community is already there.
>
> (Sabrina Cantor, Oral History)

There was a strong social life that included both on-siters and off-siters. Many people in their oral interviews mention the frequent parties.

> A party at CAT would be a meal, invariably a meal. It would involve people performing, reciting and you would just be enriched by all these different amazing people.
>
> (Graham Preston, Oral History)

These parties involved everybody, whether helping prepare the food, entertaining the children or performing in a band. These activities fostered strong bonding capital.

In addition, friendships and connections often remained long after people had left the area. Many people comment on how they made lasting friendships at CAT (including the author). Graham Preston who never lived on-site, observed:

We were a very close-knit community of people. And I have friends, all my friends, all my close friends are people who are associated with work. I built relationships with people at CAT that are extraordinary.

(Graham Preston, Oral History)

There is a canon of mythic narratives that contributes to its corporate culture, and many of these myths can be identified in the oral histories. Particularly important are the foundational myths and the accounts of significant moments, such as Gearchange. However, there are also other mythic narratives, such as the 'great pig debate', discussed below. These shared values, identifiable heroes, ritual events and mythic narratives all contribute to the bonding social capital.

Sharing

Sharing, as I have indicated above is a necessary, albeit not sufficient criteria for defining a community. The idea of sharing was central to the CAT ethos. However, sharing is a complex and contested concept. Sharing can involve anything from the idea of a common humanity to physical intimacy with a loved one. However, the idea of community is located somewhere between the universal and the intimate and suggests sharing with some people and not others. Russell Belk suggests that 'rather than distinguishing what is mine and yours sharing defines something as ours' (Belk 2014: 10).

The question arises about what must be shared, with whom and to what extent in order to constitute a community. There are five aspects of sharing that are necessary to form strong bonding social capital: locality, ideas, resources, lifestyles and responsibilities. All five of these elements must be present to some extent to constitute a community. These elements can be shared to different extents with different people within the community. Conversely, I can share one or possibly more than one of these elements with people who are not necessarily part of my community. I can share ideas and lifestyle with others, but unless there is some geographical proximity, sharing of responsibility and resources, then these others cannot be said to belong to the same community. I will discuss the first four of these aspects of sharing below but address the idea of shared responsibility in Chapter 3.

Locality

Communities are often thought to be spatial phenomena – sharing a particular locality. The parameters of what constitutes locality have extended with modern transport and communications. Howard Rheingold, in his book *The Virtual Community* first published in 1993, suggests that online interaction with geographically distant people through an early computer conferencing system WELL (Whole Earth 'Lectronic Link) 'felt like an authentic community' (Rheingold 2000: xvi). Religious traditions are sometimes characterised as communities, for example, the Hindu community. This idea of religious community extends globally and is often validated by the erroneous idea that a Hindu will be familiar, comfortable and made welcome in a temple whether located in Mumbai or London. I suggest that these geographically extended interactions, whether facilitated through digital communications or shared beliefs, constitute social networks rather than communities *per se*. Networks are looser interactions than communities and tend towards bridging social capital.

Andreas Wittel contrasts the concept of network sociality with community. Wittel suggests that community 'entails stability, coherence, embeddedness and belonging', whereas networks are relatively 'open structures, able to expand almost without limits, and they are highly dynamic'(Wittel 2001: 52). Wittel overemphasises the distinction between networks and communities as central nodes of networks can coalesce into a community and communities can diffuse into networks. CAT, for example, can be thought of as both a community and a network. CAT as a community is concentrated in the Dyfi Valley, whereas CAT as a network is geographically dispersed through the CAT members association, ex-members, ex-volunteers and students. This more diffuse association of individuals may share proximity, lifestyle and ideas, but not necessarily other facets that determine a community.

It is problematic to determine the boundaries of a specific locality. Borders are vague and permeable, especially in an age of modern transport. However, I will define the locality as the Dyfi Valley. The river Dyfi flows from just above Dinas Mawddy in the north, past Machynlleth and into the sea at Aberdyfi approximately 30 miles away. It also includes the coastal town of Aberystwyth. The Dyfi Valley itself is often referred to as a distinct place both in the recordings for the oral history project and by my interviewees. The idea of the Dyfi Valley as a distinct locality has been reinforced by its designation as a biosphere in 2009 by UNESCO and the formation of organisations such as Ecodyfi, a third sector organisation that aspires 'to foster sustainable community regeneration in the Dyfi Valley (Ecodyfi 2022). There are of course many different communities that share this locality of the Dyfi Valley that do not necessarily share the other facets. These different communities overlap and interact in complex ways.

Ideas

I deliberately use the term 'ideas' here, rather than ideology. Ideas tend to be vaguer and do not necessarily have the same sort of political connotations associated with ideology. For example, Pete Raine, the second Director, notes, 'I don't know if there was a single thing that linked everybody although everybody would have tried to be environmentally conscious' (Pete Raine, Oral History). Clive Newman, who was the site engineer from the mid-1980s until early 2000, rather disingenuously suggested: 'I would not describe myself as being an environmentalist' (Clive Newman, Oral History). However, although not informed by a clear ideology, the wider CAT community does share the idea of practical utopianism. This as Erik Wright (2010) has suggested has three aspects – identifying the harms caused by human activity, proposing viable and achievable alternatives and suggesting methods of transforming society. In the case of CAT, the harms are predominantly perceived in environmental terms and the viable alternatives are various models of a more environmentally sustainable society. An environmentally sustainable society is primarily achieved through utilising renewable energy sources that are non-polluting. This does not mean that CAT is solely about replacing fossil fuels and nuclear energy with wind and water turbines and solar power. Organic gardening, land use and reduction of meat consumption, improving home insulation and so on are all aspects of AT. However, these really practical solutions necessarily imply a vision of a better way of being that, at least to some extent, challenges accepted mores.

While not as explicitly radical as groups such as Earth First, CAT can be considered as implicitly oppositional. Bob Todd et al, in the 1977 publication *An Alternative Energy Strategy for the United Kingdom*, argued that 'the conventional belief that happiness is proportional to energy consumption would need to be replaced' (Todd et al 1977: 4). Gerard Morgan-Grenville in the position document challenges the ethos 'of our resource-hungry civilisation – the throw away attitude to consumer products' (Morgan-Grenville 1973). More recently, CAT's first Zero Carbon Britain Report (see Chapter 5) calls for 'a thorough overhaul attitudes to energy consumption' (Helweg-Larsen and Bull 2007: 7). The term 'alternative' itself implies that sustainability can no longer be 'business as usual'.

While the idea of practical utopianism is a common thread throughout CAT's history, this is not to suggest that these ideas are static. As I indicated in Chapter 1, utopianism is not an immutable ideology, but a dynamic way of thinking about the world. Consequently, while a better way of being in the early decades was informed by Schumacher's idea that 'small is beautiful', the concept of sustainability and the aspiration to achieve a carbon-free society now resides at the heart of CAT's thinking. The official CAT guidebook from 1991 categorically states:

> If we want to survive into the future – and have a relatively smooth ride – our best bet lies with understanding and working with natural processes, rather than trying to conquer nature.
>
> (CAT 1991)

Everyone who has ever worked at CAT from the very early pioneers to the current senior management team would concur with this idea. This same idea is now couched in slightly different terms as CAT's vision is expressed as 'a sustainable future for all humanity as part of a thriving natural world' (CAT 2022).

Resources

Clearly, the site community shared resources. One of the advantages of living on-site was that although salaries were very low, any money left after paying a small rent and a contribution for food was available for small luxuries. Often, for those who had not got on the housing ladder, did not have any substantial savings or did not have a partner earning a more realistic salary, there was little option other than living on-site. Nonetheless, all the people who lived on-site actively chose to do so, and living on-site was integral to their experience of being at CAT. The main resources shared were food, the communal space of Tea Chest and electricity.

Including volunteers, there were between 20 and 25 people living on-site in the 1980s. This enabled quite a saving by bulk buying food, particularly wholefoods. Most members of the site community would gather in Tea Chest for a shared evening meal. There was a rota for cooking for the weekdays. Many commented that this was a mixed blessing. On the one hand, you were guaranteed an evening meal during the week and you only had to cook once every two or three weeks. On the downside, when you did have to cook, you often had to cook for 20 people, which was quite daunting for some people. In addition, some people were inevitably better cooks than others, so the quality of meals was always quite variable. There was also a site community rota for baking bread. Again, the quality was variable – the brick-like loaves that some people produced could hang around for days, whereas a good bake could disappear in a couple of hours. The cottages and the caravans also had some basic cooking facilities so people, if they chose, could prepare a simple meal for themselves. The ingredients and basics such as tea and coffee were all available from the community food store.

It was not just the site community that shared food. During the week, there was a shared midday meal in Tea Chest. There was also a rota, which included both on-site and off-site staff, for cooking these lunches. This meant that everybody had to cook lunch at some point. Again, there was the benefit of only having to cook occasionally and having a daily cooked meal. However, this could involve cooking for 35 to 40 people and this inevitably took quite a considerable amount of time from your working day. This was occasionally resented by some who felt, for example, that fixing a broken wind turbine was more important and a better use of their time.

> A few people felt 'I am far too important a person and I have qualifications. Why am I sitting here chopping carrots?'. They could be quite uppity about this.
>
> (John, interview February 2021)

Nonetheless, the shared lunches were an important arena for the formation of bonding capital. Tea Chest lunches were where you could catch up with people whom you did not interact with in your daily work, where you could discuss work-related problems, debate the cultural, political and environmental issues of the day or simply chat. This drew together on-site, off-site workers and volunteers, creating a sense of community.

One other aspect of sharing resources is also worth mentioning. From the early 1980s to 1990s, as part of the self-sufficiency ethos of CAT, there was a smallholding at one end of the site with ducks, geese and chickens, and in the very early days a cow as well. For a number of years, CAT raised two pigs. This made good sense, because the pigs could be fed scraps and vegetable trimmings from the restaurant and Tea Chest. At the end of the season, someone from the local abattoir came on-site to kill and butcher the pigs. It was often wrongly thought that everyone at CAT was vegetarian, as both the Quarry Café in Machynlleth and the on-site restaurant did not serve meat or fish. The meat from the pigs was frozen and was available for all workers on- and off-site. Many of the meat eaters not only waxed lyrical about the quality of the pork, but also indicated that they also felt it was morally responsible to witness the killing of the pigs if they were going to eat the meat.

The raising, slaughtering and eating of the pigs was important for a number of reasons. Primarily, it was about sharing resources. However, it also demonstrated some of the debates, not only at CAT, but within the wider green movement. From one perspective, raising and eating pigs in mid-Wales makes perfect sense. It is consistent with both self-sufficiency and bioregionalism. Bioregionalists argue that it is important to adapt to the ecosystems of one's immediate environment and live in harmony with the resources of one's locality as far as possible. Using kitchen scraps to feed pigs makes good environmental sense. In addition, the area around the Quarry is not good arable land, and therefore it is logical that the diet should contain some meat. However, many members of CAT were vegetarian or vegan and many (albeit not all) objected to the killing of animals. The vegetarians and vegans at CAT refrained from eating meat for a wide variety of reasons – political, moral, personal well-being or a combination of these. There is also an environmental argument for vegetarianism and veganism. Animal rearing is very resource intensive. The ever-increasing global demand for meat has led to the destruction of many natural regions and therefore contributed to the loss of biodiversity. Intensive animal farming also produces significant greenhouse gas emissions. Each year, the question of raising and killing of the pigs was debated. However, one year, the debate was particularly extended and heated, and this has become generally known as 'the great pig debate'. Many people at CAT have heard of this 'great pig debate', even if they were not around at the time, and it has become a mythic narrative that is part of the fabric of a community memory. In other words, a community not only shares tangible resources, but it also shares less tangible resources such as stories and memories.

John Urry in his oral interview gives the fullest account of 'the great pig debate'. In his account, one of the long-term volunteers was a particularly passionate vegan. She had also been responsible for looking after the pigs. When the time came, she and some others were particularly adamant in objecting to the pigs being killed. This meant that the meetings in which this was discussed came to a deadlock and no consensus could be arrived at.

> We went through at least two meetings, now just think of the total staff time that had then gone into this discussion … It became quite an issue.
>
> (John Urry, Oral History)

A way through this impasse was suggested by a member of staff who bizarrely suggested that the pigs themselves should be consulted. The idea that he proposed was to get a medium to communicate with the pigs. It is probably an indication of the frustration that everyone on both sides of the debate felt that this was agreed to. A medium was found, who claimed to have communicated with the animals. The medium reported that the male pig was an old soul and was keen to get back on the wheel of transmigration and so was not particularly concerned about losing his life. The female pig on the other hand communicated to the medium that she would like to raise a litter, so did not wish to be killed just yet. The outcome was that the male pig was killed and butchered, and a good home was found for the female pig, satisfying all. It is not important whether the member of staff who made the suggestion actually believed that the pigs could contribute to the debate about their demise or whether this was simply a canny political move to end the stalemate. It is also not important whether anyone else believed that the pigs could communicate. What is important is that it reveals something about the decision-making processes (see Chapter 3). More significantly, it has become part of the narrative of the community – irrespective of whether individuals believe it to be preposterous or not.

There were two other aspects of shared resources that contributed to creating strong bonding social capital – a limited energy supply (discussed below) and a relative flat pay structure. Members of CAT were paid a very basic salary much below what many could have earned working elsewhere. However, perhaps the most radical aspect of CAT was that there was a relatively flat pay structure – the Director, the engineers and administrative staff all got paid the same. This equal pay structure produced 'an intense collective loyalty' and a sense of 'we are all in this together' (Peter Harper, Oral History). This collective attitude was exemplified when in the early 1990s there was a cash flow problem and everyone who could afford agreed to have their salaries suspended for a couple of months.

Lifestyle

Lifestyle is a very nebulous concept. However, as a starting point, David Chaney suggests that 'lifestyles are patterns of actions that differentiate people' (Chaney 1996: 4). These patterns of actions manifest in daily life. As already indicated, many outsiders believed that members of CAT had a distinctive lifestyle that was associated with the counterculture and encapsulated by the label 'The Shit and Wind'. However, most members of CAT had no association with the counterculture and were probably as far removed from the hippie lifestyle as merchant bankers. In other words, the perceived differences of the lifestyle set CAT apart from others in the area and represented CAT as a distinct community.

The lifestyle was informed by the broad green ideas that people at CAT shared, the relatively low income that they received and the rural location in which they lived. There was not a marked difference between on- and off-site lifestyles. While it was much easier to maintain a low environmental impact living on-site, all off-siters made an attempt to live in as environmentally friendly way as possible. This does not mean that patterns of actions were homogenous or that cultural tastes were uniform. There were also debates and differences, for example, on the use of cars. Whether people at CAT have a distinctive lifestyle that differentiates them from others is more a perception of outsiders than a reality. In the very early days, a few of the pioneers had a more countercultural look, and occasionally, a long-term volunteer would have dreadlocks that made them clearly identifiable, but this was by no means true for the majority. Some of the everyday practices, such as recycling and using organic fair-trade products as much as possible, were perceived as radical by some but are now well incorporated into mainstream lifestyles. Consequently, lifestyle choices are increasingly shared beyond the CAT community. A number of the local Welsh community who were interviewed for the oral history project note that they thought CAT did have some impact on the wider community of the Dyfi Valley. Meri Wells comments that 'undoubtedly, the area is more eco-friendly than it might well have been had it not been for CAT being here'. She observes that the local shops tend to stock more organic and fair-trade products than did other comparable towns in mid-Wales. It is impossible to actually establish whether there is a direct causal link between the presence of CAT and changes in lifestyle choices of other people in the Dyfi Valley. Nonetheless, it is significant that CAT lifestyles are no longer so clearly differentiated from those of other local people, and this is both a product of greater integration and has enabled CAT's agenda to be more widely accepted in the Dyfi Valley.

Self-Sufficiency

As there were both on- and off-site members, there was an ongoing debate about the purpose of the site community and whether it was integral to CAT's mission. In the very early pioneering days, everybody lived on-site and developing a community who lived with AT was perceived as the primary raison d'être of the project. It is clear that Gerard Morgan-Grenville intended that an on-site community living a low environmental impact lifestyle would be a significant aspect of the embryonic CAT project. He suggests in his memoir that after his visit to the States, probably inspired by places like The New Alchemy Institute, he started to consider setting up:

> A centre where the need for a simpler less damaging lifestyles could be explained and demonstrated to the public. The aim of greater self-sufficiency was also to be a principle of the project … To gain public credibility those who lived and worked there would be as self-reliant as possible.
>
> (Morgan-Grenville 2001: 158–9)

Morgan-Grenville alludes to the concept of self-sufficiency, which was a significant trend in the green movement of the 1970s. While self-sufficiency was not necessarily associated with an Arcadian aspiration, it was inherently critical of industrialisation and therefore was often orientated to a more rural way of life that is epitomised in John Seymour's 1976 publication *The Complete Book of Self-Sufficiency*. This is an expanded version of *Self-Sufficiency*, which he co-wrote with his wife and was published in 1973. The *Complete Book of Self-Sufficiency* sold well and has been reissued several times. The latest edition published in 2019, with a foreword by the TV chef and presenter Hugh Fearnley-Whittingstall, now has the subtitle *The Classic Guide for Realists and Dreamers*.

John Seymour and his wife ran a 5-acre smallholding in Suffolk in the late 1950s and then moved to a small farm in Pembrokeshire in 1964. Seymour emphasises that self-sufficiency is not reverting to a primitive austerity but moving forward 'to a new and better sort of life' (Seymour 1976: 7). In this beautifully illustrated book, it is clear that *The Complete Book of Self-Sufficiency* is intended as a guide for those moving from an urban context to a more rural environment, but who have not necessarily got sufficient skills. There are short sections in this book on everything from animal husbandry to making willow baskets.

There are three significant and interconnected aspects to Seymour's book: independence, rural living and ecologism. Self-sufficiency clearly suggests being as reliant as possible on one's own skills and labour, and not being dependent on the products of industrialisation. This is of course a challenge, and there was an ongoing debate at CAT as to how far one could actually be self-sufficient, and where it might be necessary to compromise for the sake of expediency. While John Seymour suggests that his book is not intended only for a rural living, and there are, for example, recipes for making jam that can easily be done in an urban kitchen, the majority of the book pertains to the skills that relate to life in the country. This can be related to the back-to-the-land movement. There was not as large a back-to-the-land movement in the UK in the late 1960s and early 1970s as in the USA. Nonetheless, there was a stream of the British counterculture that perceived 'a base in the country, almost by default, was seen as providing a degree of distance from mainstream straight society' (Halfacree 2006: 316). It is unclear whether Steve Boulter, whom Gerard Morgan-Grenville had commissioned to find a site, only looked at rural locations. However, the disused slate quarry, which was a serendipitous find, was for many reasons ideal. However, its rural location did raise the criticism that the ideas promoted by CAT were not relevant for urban living. However, CAT did try and address this criticism in several ways, for example, showing that it was possible to grow vegetables in a small urban garden. Seymour's book clearly was underpinned by the ideas about ecologism that were emerging in the 1970s. Seymour suggests that 'Man should be a husbandman, not an exploiter. This planet is not exclusively for our own use' (Seymour 1976: 7). In addition, Seymour has a section towards the end that briefly discusses renewable energy.

CAT in its early days had a smallholding with goats, pigs, geese and chickens.

> Many years ago, when we started, this whole area was very much more to do with self-sufficiency because of course back in the year dot what we all wanted to do was to live off the land, close our doors to the wicked outside world and become a sort of little self-sufficient island. That was I think very much part of the early ethos of the place.
>
> (John Urry, Oral History)

In one sense, the site was an extremely good choice for the embryonic project. It had several derelict buildings that could be renovated and a reservoir that was easily utilised for powering a water turbine. However, in another sense, the disused quarry was totally unsuitable for attempting any degree of self-sufficiency. The site, which is predominantly composed of the waste from slate quarrying, was never going to be able to produce sufficient food for even a small pioneering community. In addition, the idea of self-sufficient communities quickly fell out of favour as it became increasingly apparent that it was a pipe dream. Self-sufficiency is a myth and, as Phillip Vannini and Jonathan Taggart observe in their ethnography of off-gridders in Canada, 'pragmatic compromises' with ideals were always necessary (Vannini and Taggart 2015: 167). The question that this raises for all people concerned with the human impact on the environment is what aspects of human activity can be retained and what has to be abandoned or changed in order to avert ecological disaster. Where does the balance lie between pragmatism and idealism?

Furthermore, many in the green movement, including those at CAT, realised that environmental destruction was not going to simply be solved at the level of personal lifestyle, but was also a political, social and economic problem.

> The whole ethos of alternative technology, which I guess at the time was very much focused on the concept of self-sufficiency, which for all sorts of reasons I think is a load of rubbish now. But it was incredibly appealing to a lot of people in the 70s.
>
> (Roger Kelly, Oral History)

The other point that Urry makes in the quote above is that the aspiration to create a relatively self-sufficient community also involved excluding 'the wicked outside world'. This ethos was informed by the apocalyptic vision of the 1970s environmental movement that is evoked in publications such as *The Limits to Growth* and *A Blueprint for Survival* (published by *The Ecologist* in 1972). This apocalyptic fear suggested that it was necessary to learn how to adapt to the almost inevitable collapse of civilisation caused by the environmental devastation of industrialisation.

Peter Harper perceived the early CAT as a sort of refugium and compared it to the monks of Lindisfarne – keeping an island of civilisation and skills alive whilst all else degenerates into an ecological apocalypse.

> At that time, there was a very widespread apocalyptic feeling about the imminent collapse – either because of a nuclear war or ecological collapse. Rather similar to today in lots of ways. So, under those circumstances, what do you do if you think there's going to be a great big collapse? Well, you might well start a place that knows how to rebuild after a collapse ... You would need centres where skills are preserved and new methods of doing things. How to rebuild on the basis of very little and very simple principles and technologies.
>
> (Peter Harper, Oral History)

This inward turning attitude indicated by John Urry and Peter Harper to the perceived looming collapse can be called the hermitic response. However, others had a more outward looking perspective that can be termed the prophetic response. This prophetic response is characterised not by creating a sort of refugium, but by the imperative to inform others of viable solutions to the environmental devastation caused by much human activity.

The distinction between the hermitic and prophetic responses to the environmental damage caused by human activity reflects another aspect of the debate that can be termed adaptation or mitigation. The hermitic response tends to be more pessimistic. This perspective suggests there is no way to completely turn around the destructive trajectories indicated in *The Limits to Growth* and therefore we have to learn to adapt to the inevitable degeneration of the environment. Mitigation on the other hand tends to be more associated with a prophetic perspective. Mitigation suggests that there are ways to reverse the destructive trends of human activity; consequently, it is vital to persuade others to adopt more sustainable patterns of behaviour. Adaptation tends to focus on individual or small-scale responses to environmental issues. Mitigation, while it does not ignore the individual, has a greater potential to incorporate social, economic and political responses. Once again, it is important to see adaptation and mitigation as a continuum of attitudes rather than polar opposites. To a certain extent, this debate between adaptation or mitigation receded from the 1980s to 1990s. However, it has once again become prevalent in the debates about how we should respond to climate change (see Pfister, Schweighofer and Reichel 2016: 39–42).

Clearly, the prophetic response won the argument as after the discussion between Rod James and Gerard Morgan-Grenville, CAT opened as a visitor centre in 1975 and shortly after began running short courses on various aspects of AT. This raised the ongoing debate on the purpose of the site community and the relationship between the site community and the wider educational aspect of the CAT project. Was the site community simply providing cheap accommodation for workers or was it integral to the message that CAT was promulgated?

The Site Community: Integral or Peripheral?

It is clear from the writing of Gerard Morgan-Grenville that the community side of CAT was an essential aspect of his vision. Morgan-Grenville perceived the site community as both an experiment and a demonstration of AT. The Centre was intended to be more than simply testing whether the technology *per se* was efficient. It was also intended as an experiment in living. There was no point in simply raising the drawbridge; it was important to demonstrate to the public that leading a sustainable lifestyle was possible. The site community therefore could be considered as being integral to the educational and communication aspect of the display. The site community was considered to have the potential to dem-onstrate that a sustainable lifestyle is possible. The site community in many ways validated the CAT project in the early days, providing a coherence to CAT's message and serving as a potential draw for visitors.

> Visitors didn't just want to come and see bits of technology or displays. They wanted to know whether and how people could live their lives using those technologies.
>
> (Roger Kelly, Oral History)

Being regarded as integral to the educational aspect of CAT was an ongoing challenge to the site community, who sometimes felt that their personal and private lives were to some extent on display. Liz Todd who initially lived on-site with her husband Bob and young family explained:

> It wasn't an easy place for young children to be. For a start, they were very much on show all the time. They were observed as a sort of almost an interesting exhibit.
>
> (Liz Todd, Oral History)

A number of people including Peter Harper and Roger Kelly, both of whom lived on-site, believed that, as the primary aim of CAT was to demonstrate the possibility of living a low environmental impact lifestyle, the site community was integral to the overall CAT project. Others suggested that living on-site was a practical and/or social decision. There were two aspects to the more prag-matic decision to live on-site. The first was purely for financial reasons; it was cheaper to live on-site and many could not afford mortgages or off-site rent. The second is that it was much easier to live a low environmental impact life-style living on-site as resources could be shared, power was primarily derived from renewable energy and there was no necessity to drive to work. There were always people on-site to socially interact with, whether simply for company or to provide a readily available pool of trustworthy child-minders. There was in addition the need for at least one on-site engineer who could respond to any emergency, such as a power outage during the night, particularly if there was a residential course.

All the people who had lived on-site not only valued the social aspect, but also valued their own personal space.

> I liked having my own space. The fact that I had my own completely sep-
> arate living quarters, but there was also a community building where we
> could eat together. If I chose to socialise there were people very near that
> I could go to their houses or they would come to my caravan was an ideal
> situation for me.
>
> (Annie Lowmass, Oral History)

However, Roger Kelly believed that the site community was more than simply about providing accommodation and a social life, and it should have been incorporated much more comprehensively into the display:

> I was convinced there had to be some way that the visiting public could
> be more involved in the community. And I even got to a point where I
> thought that maybe the solution was that it should be a part of the contract
> for living on-site. That to some extent you had to allow the public to come
> in and you had to talk to them about the way you lived your life.
>
> (Roger Kelly, Oral History)

Kelly acknowledges that being on display would be difficult and that living on-site should be for a limited period only.

Peter Harper, who lived on-site for 12 years, expressed his disappointment with the site community. He observes that not only should the site community demonstrate sustainable living but it also had a responsibility to measure and record what environmental impact living in this experimental community had.

> There is something unique here. We have a unique site, renewable energy.
> We have got a unique community and we are supposed to be demon-
> strating sustainable living ... We should measure and record what we do;
> start car clubs, measure our carbon emissions, all that sort of thing in order
> to show this is a model of how it could happen. And it could be reproduced
> elsewhere.
>
> (Peter Harper, Oral History)

However, both Kelly and Harper were lone voices in suggesting that the community must serve CAT's wider educational mission. Most site community members tended to live on-site for more personal and individual reasons.

Living Off-Grid

Until 1990, CAT was entirely off-grid and primarily relied on renewable energy for all its electric requirements. This meant that electricity was a finite resource that also had to be shared, and this also contributed to creating social bonding capital. Renewable energy is in many ways the central core of the CAT project. In the 1970s and 1980s, renewable energy tended to be small scale and often rather DIY; this is exemplified by the wood beamed and sail cloth Cretan windmill that used to be located on a prominent ridge. This never actually generated much power but became symbolic of the early CAT ethos. Although wind power tended to represent renewables as wind turbines are clearly visible and have a certain aesthetic appeal, hydropower generated most of CAT's electricity. In the 1970s and 1980s, solar energy had not developed sufficiently to generate much in the way of power. Early photovoltaic cells were both expensive and inefficient. An early display of photovoltaics, which although made an interesting exhibit, only generated about 50 watts.

One aspect of the site was to show that it was possible to live and work with renewables. Consequently, being off-grid was an important signifier of CAT's mission. This was symbolised by a green line on the path that visitors took from the car park to the site. Bob Todd, whose expertise was absolutely critical in installing CAT's energy system, observed that:

> One of the early things when we were discussing the energy system was the idea of trying to make the place self-sufficient in energy. To prove to people it could be done. Having the electricity system completely independent did help credibility in the early days.
>
> (Bob Todd, Oral History)

Todd does add that the site was not totally self-sufficient for power. Heating and cooking were significant consumers of energy. Most of the heating was generated by wood burning stoves and even at one point coal. Direct solar was used in an experimental inter-seasonal heat storage system, which was one of the first in the UK, to heat the main office. It was not as successful as hoped but did provide background heating all year round. However, all electricity for the site was generated by renewables for the majority of the time. As there was a visitor circuit, which included a restaurant,[3] the public had to have sufficient facilities, so there was an emergency backup diesel generator. Overall, the site ran on about two and a half kilowatts, which had to be shared.

> You learnt the value of the fact that it was those first few hundred watts that changed your life because they're the ones that enable you to have light at night.
>
> (Pete Raine, Oral History)

There was often a call on the intercom from one of the families wondering if there was sufficient power to run the washing machine, and often the lights would dim when someone turned on a power-hungry machine. In very dry years, when there was little water in the reservoir; the site sometimes had no power at night and was pitch black.

Comfort and Convenience

Having a very limited and somewhat intermittent supply of electricity is very different from most people's everyday experience. Electricity, except for very rare instances in the UK and other parts of the developed world, is always available at the flick of a switch. This ready supply of power to light and heat our homes and power our gadgets is taken for granted. On-grid homes are represented in terms of convenience and comfort, suggesting that everyday life in off-grid homes is relatively inconvenient and less comfortable. In contemporary discourses, convenience and comfort are valued and linked to the idea of progress. Vannini and Taggart suggest that 'convenience is now synonymous with lack of complications and with a lifestyle rendered easy by countless consumer products and services' (Vannini and Taggart 2015: 107).

It is of note that developments in technology are frequently associated with convenience. Thomas F. Tierney suggests that 'the value of technology in modernity is centered on technology's ability to provide convenience' (Tierney 1993: 6). Tierney defines convenience as 'the ability to mitigate the effect of bodily limits' (Tierney 1993: 38–9). These bodily limits are primarily spatial and temporal. Hence, convenience is articulated in terms of satisfying needs easily and quickly. This is nicely illustrated by Hew Jones, a local farmer and parish councillor. He noted that there were many small-scale hydro systems in the area. He himself had a water turbine that supplied limited and intermittent electric power to 11 neighbouring properties. One of the things that he noted was that leaves often blocked the sluices, and he recalls an occasion when 'I had to get up from the chapel in the middle of a service because the light was getting dim and go out and rake the leaves'(Hew Jones, Oral History). In 1961, the area was connected to the national grid and there was a reliable, cheap and easily accessible electricity supply. Jones notes that in 1974 when CAT started, locals were somewhat bewildered by 'a lot of screwballs who were talking about putting back water turbines' (Hew Jones, Oral History). Why would anyone want to actively choose something that is palpably less convenient when a more convenient option is readily available? As Tierney observes, the concept of convenience is 'deeply ingrained in modern culture' (Tierney 1993: 10) and 'guides the consumption choices of modern individuals' (Tierney 1993: 13). Anything that is perceived as less convenient to other options is conceived as being regressive. Not being connected to the grid, as Bob Todd observed, gave CAT a certain credibility but the perception that this was less convenient and retrogressive was also barrier to the acceptance of renewable energy.

Tierney also indicates that convenience has the connotation of comfort, which in the context of modernity is viewed as 'a state of physical and material well-being, with freedom from pain and trouble, and satisfaction of bodily needs' (Tierney 1993: 40). Comfort is perceived as a relationship between the body and the physical environment. As a somatic concept, comfort has a number of facets connected to our senses. These facets include cleanliness, sight, hearing and heat. If the physical context is regarded as too dirty, too dark, too loud or too cold, then the context is deemed to be both physically and emotionally uncomfortable. We accept that comfort is to a certain degree an individual and contextual phenomenon. The level that I play my music might be perceived as being uncomfortable for my wife, or the loudness of music at a concert would be uncomfortable in my home. However, as comfort is primarily perceived to be determined by the physical attributes of an object, it is regarded as being an objective quality. If I perceive my chair to be uncomfortable, all I have to do is to exchange it for another chair that is more comfortable for me. The facets of comfort exist along a spectrum. At the extreme ends of the spectrum, the objective physicality does become decisive. For example, extreme heat or cold will ultimately be universally fatal for all humans. However, in the mid-range of the spectrum, comfort is culturally determined. Think, for example, about the range of light and warmth that might have been found comfortable in the Middle Ages as opposed to now.

The perception that comfort inheres in the physical context poses a challenge for trying to promote a more sustainable lifestyle. It is clear that the limited electric supply at CAT did have an impact on the levels of light available to the site community. Vannini and Taggart observe that 'visual comfort, in a nutshell is malleable, yet it is often perceived as natural and fixed as if an optimal standard could be positively measured, assessed and uniformly agreed upon' (Vannini and Taggart 2015: 95). Consequently, if levels of light are lower than 'the optimal standard', illumination is perceived as being dim and therefore gloomy and uncomfortable. However, while many of the site community in the oral history interviews mention the limited availability of electricity and concomitant low level of lighting, no one articulated this in terms that living on-site was uncomfortable. Vannini and Taggart call this facility to find comfort outside the normative parameters 'The Thoreau Effect'. There are two aspects to the Thoreau Effect. First, to find comfort in what other people might consider as 'inadequate, obsolete, cumbersome, inferior, inconvenient and uncomfortable'. Second, to appreciate 'comfort obtained through hard work and direct physical involvement' (Vannini and Taggart 2015: 98). They contrast this with what has been called the Diderot Effect after the French philosopher Denis Diderot.

> If the Diderot effect works somewhat like an addictive search for greater doses of comfort obtained through deceasing effort and inconvenience, the Thoreau Effect works by way of reversing that trend. It is a phenomenon marked by contentment, rather than perennial dissatisfaction, with the affective capacity for comfort from whatever one already has.
>
> (Vannini and Taggart 2015: 97)

There are two interrelated challenges for those, like CAT, that are trying to promote a sustainable lifestyle. First, there is a perception that a sustainable lifestyle is inherently less comfortable than the standards of living to which we are generally accustomed. Second, an attitude of finding comfort with what we already have challenges the dominant values of consumer society, which constantly ratchets up our expectations that the new supersedes the old and will invariably make life more convenient and comfortable.

Consequently, there is a perception that the AT lifestyle is intrinsically uncomfortable. What was considered to be a luxury is now thought of as being a necessity for individual well-being. This is a challenge not only for CAT, but also for the environmental movement more generally. For example, owning a car, particularly in a rural area, is now perceived as a necessity rather than a luxury. However, although there is an increasing move toward electric vehicles, simply replacing cars that run on fossil fuels with electric cars is not the panacea for addressing the environmental crisis of our times. Dennis Kingsley and John Urry (2009) argue that the whole system of the car must be changed to achieve low carbon lives, including the idea of private ownership of vehicles. Yet car ownership is not only about perceived convenience or rational decisions, 'it is as much about aesthetic, emotional and sensory responses to driving, as well as patterns of kinship, sociability, habitation and work' (Urry 2011: 132).

The Demise of the Site Community

The site community was eventually wound up in 2010. There were a number of intersecting reasons for this, both internal and contextual. The increase in staff numbers reached 150 in 2005 (Harper no date: 23). This meant that there were considerably more people living off-site than at the Quarry itself, which seemed to make the site community less important. The purpose of the site community was increasingly questioned. This scepticism about the validity of having an on-site community was exacerbated by the shift away from the ethos of self-sufficiency.

> By the mid-90s, it was becoming obvious there was no point in CAT going on about what a handful of people could do living in an abandoned slate Quarry. It was no longer relevant to the world … It is difficult really to justify it as anything other than, oh, it's nice to have a few people living on-site.
>
> (John, interview February 2021)

Additionally, fewer people wanted to actually live on-site, and eventually there was only one family living at the Quarry.

There were two other practical issues that led to the final demise of the site community. First, Tea Chest was forced to close because of health and safety. A decision was made that with the increasing number of staff to feed, and cooking communal meals was increasingly challenging and time consuming, the restaurant staff would cook the staff lunches. However, the unforeseen consequence of this decision was Tea Chest was no longer classified as a private kitchen, but as an extension of the restaurant and therefore liable to inspection by public health officials. Tea Chest is an old building and it was almost impossible to maintain the standards required by public health without major investment.

Consequently, when it was visited by an inspector, it did not meet the requisite standards and was closed. The second reason was that CAT as the legal owner of the site and its assets was effectively a landlord. While people were quite content with the somewhat basic standard of living on-site, by 2010 the accommodation no longer met legal requirements. There was always an issue about maintaining property on-site. The priorities of builders and engineers inevitably tended to focus on maintaining or developing the public areas. Finances were always tight at CAT and when there were any available funds, these were used for maintaining or developing the visitor circuit, rather than investing in the personal accommodation. In many ways, the community was a victim of CAT's propensity to be too many things. As a self-contained intentional community with members intent on living as sustainably as possible, it might well have survived. However, this was never CAT's raison d'etre. As their current maxim suggests, 'CAT's mission is to inspire, inform and enable humanity to respond to the climate and biodiversity emergency'. A small group of people, no matter how sustainably they live, has no longer got the capacity to inspire, inform or enable a wider public to adopt more environmentally friendly ways of being.

From Community to Network

Community was an integral aspect of the CAT project at the beginning. The community was not only intended as a living experiment in sustainability, but also to be a demonstration of living a low environmental impact lifestyle. In many ways, the site community at CAT was a precursor of ecovillages, which as Jonathan Dawson observes, have moved from 'relatively isolated countercultural experiments offering a profoundly alternative vision and lifestyle to the cultural mainstream to increasingly working in formal and informal alliance with more progressive elements in today's society'. Dawson succinctly suggests in the title of his essay that ecovillages have transformed from being islands to networks. This shift from 'islands' to networks has been primarily because of changes in wider society (Dawson 2015: 217–18).

Dawson suggests that CAT was not only a founding member of the Global Ecovillage Network (GEN) but is also an exemplar of an ecovillage (Dawson 2015: 225).[4] CAT, except in the very early pioneering days, was never an island as it opened its doors to the general public in 1975. However, as CAT developed its networks, the site community became less significant. CAT is, as many people comment, a very different organisation from what it was. There is a sort of nostalgia, and some even suggest that CAT has lost something because of the demise of the site community. However, most people acknowledge that, since the waning of the idea of self-sufficiency and the attenuation of the small is beautiful ethos, the site community became increasingly irrelevant. The site community was a bold, and radical experiment in trying to find practical solutions to environmental issues. The environmental issues that threaten life on planet earth are now widely acknowledged and CAT's strategy has adapted reflect this changing context. Nonetheless, the formation of the strong bonding social capital was a necessary foundation to forming effective bridging social capital and enabled CAT to develop its networks of influence, which have not only ensured CAT's survival through difficult times but have also contributed to its credibility.

Notes

1 A fourth caravan was moved on-site for me in 1984. It was considered beneficial to have someone who was running the restaurant on-site as most of the courses required providing evening meals and breakfasts. All of these caravans were got rid of in 1990, as they were located in a part of the site that was incorporated into the public area as part of the Gearchange project.
2 In the summertime, when CAT was very busy, local people were often employed on a casual basis to help run the restaurant, bookshop and so on. While some of these casual staff might well be committed to environmental issues, have very fond memories and be very loyal to CAT, they were not involved in CAT in the same way as permanent staff members.
3 When I arrived to run the restaurant in 1984, I had to negotiate having a food processor, which I thought was essential for providing an interesting and varied vegetarian menu. I was eventually allowed a domestic processor, as there was insufficient power for an industrial one. Quite often, there was insufficient power to even run the domestic one, and I always had to check with the engineers before I could use it.
4 However, CAT is not listed on the current pages of GEN Europe website.

References

Belk, R. (2014). Sharing Versus Pseudo-Sharing in Web 2.0. *Anthropologist* 18(1), 7–23. doi: 10.1080/09720073.2014.11891518

Centre for Alternative Technology (CAT). (1991). *Official Guidebook*. Machynlleth: CAT Publications.

Centre for Alternative Technology (CAT). (2010). *Memorandum of Association*. Available at: https://find-and-update.company-information.service.gov.uk/company/01090006 (Accessed 26 January 2021).

Centre for Alternative Technology (CAT). (2022). *Strategy and Governance*. Available at: https://cat.org.uk/strategy-and-governance-2 (Accessed 9 January 2022).

Chaney, D. (1996). *Lifestyles*. London: Routledge.

Dawson, J. (2015). From Islands to Networks: The History and Future of the Ecovillage Movement. In J. Lokeyer & J.R. Veteto (Eds.), *Environmental Anthropology Engaging Ecotopias*. New York: Berghann, 217–234.

Deal, T.E. & Kennedy, A.A. (1982). *Corporate Cultures: The Rites and Rituals of Corporate Life*. Reading, MA: Addison-Wesley Publishing.

Delaney, G. (2018). *Community: Key Ideas*. London: Routledge.

Ecodyfi. (2022). *History of Ecodyfi*. Available at: www.ecodyfi.wales/history (Accessed 7 January 2022).

Halfacree, K. (2006). From Dropping Out to Leading On? British Counter-Cultural Back-to-the-Land in a Changing Rurality. *Progress in Human Geography* 30(3), 309–36. doi: 10.1191/0309132506ph609oa

Hands, J. (2016). *Housing Co-Operatives*. London: Castleton.

Harper, P. (no date). Demanding the Impossible: Managing and Mismanaging a Radical Agenda: A Life-Story of the Centre for Alternative Technology. (Unpublished paper provided by the author).

Helweg-Larsen, T. & Bull, J. (2007). *Zero Carbon Britain: An Alternative Energy Strategy*. Machynlleth: CAT Publications.

Kingsley, D. & Urry, J. (2009). *After The Car*. Cambridge: Polity Press.

Miller, T. (1999). *The 60s Communes: Hippies and Beyond*. New York: Syracuse University Press.

Morgan-Grenville, G. (1973). *The Society for Environmental Improvement. The National Centre for Alternative Technology*. Unpublished paper. Available in the CAT Archives at the National Library of Wales (Box 1/12).

Morgan-Grenville, G. (2001). *Breaking Free*. Bridport: Milton Mill Publishing.

Pepper, D. (1991). *Communes and the Green Vision: Counterculture, Lifestyle and the New Age*. London: Green Print.

Pfister, T., Schweighofer, M. & Reichel, A. (2016). *Sustainability*. London: Routledge.

Putnam, R.D. (2000). *Bowling Alone: The Collapse and Revival of American Community*. New York: Simon & Schuster Paperbacks.

Rheingold, H. (2000). *The Virtual Community: Homesteading on the Electronic Frontier*. Cambridge, MA: MIT Press.

Ruiu, M.L. (2016). The Social Capital of Cohousing Communities. *Sociology* 50(2), 400–15. doi: 10.1177/0038038515573473

Seymour, J. (1976). *The Complete Book of Self-Sufficiency*. London: Faber.

Shenker, B. (2011). *Intentional Communities: Ideology and Alienation in Communal Societies*. London: Routledge.

The Ecologist. (1972). *A Blueprint for Survival*. Harmondsworth: Penguin Book.

Tierney, T.F. (1993). *A Genealogy of Technical Culture: The Value of Convenience*. Albany: State University of New York Press.

Todd, R.W. & Alty, C.J.N. (Eds.) (1977). *An Alternative Energy Strategy for the United Kingdom*. Machynlleth: CAT Publications.

UK Cohousing. (2021). About Cohousing. Available at: https://cohousing.org.uk/about/about-cohousing (Accessed 26 January 2021).

Urry, J. (2011). *Climate Change and Society*. Cambridge: Polity Press.

Vannini, P. & Taggart, J. (2015). *Off the Grid: Re-Assembling Domestic Life*. Oxon: Routledge.

Williams, R. (1988). *Keywords: A Vocabulary of Culture and Society*. London: Fontana.

Wittel, A. (2001). Towards a Networked Sociality. *Theory, Culture and Society* 18(6), 51–76. doi: 10.1177/026327601018006003

Wright, E.O. (2010). *Envisioning Real Utopias*. London: Verso.

3 Decision Making

Introduction

In this chapter, I will explore the consensus decision-making processes and the relatively flat wage structure at the Centre for Alternative Technology (CAT) and consider how integral these were to the overall project. This continues the theme of sharing that I addressed in Chapter 3. Consensus decision making is indicative of sharing responsibilities and contributes to the formation of social bonding. Consensus decision making might be considered as one of the most utopian aspects of CAT. I will also outline the reasons as to why the consensus decision making ultimately failed, and whether the adoption of a more hierarchical managerial structure constituted the demise of practical utopianism.

In this chapter, I address two interlinked themes: the link between sustainability and egalitarianism and whether consensus decision making facilitated or was an obstacle to identifying practical solutions. Borgnäs et al (2015: 3) argue that environmental sustainability is 'a field that necessitates re-thinking and restructuring on most social, economic, political, technological and human levels'. Consequently, sustainability includes rethinking the way in which organisations are run, the nature of work and remuneration for labour. The internal organisational dynamics of CAT suggests a better way of being that has the potential to achieve a more environmentally sustainable and just society. The consensus decision making and relatively flat pay structure at CAT can be considered as what Darcy L. Leach calls a prefigurative style of politics. Leach (2016: 42) defines this as: 'A belief that movements can only accomplish radical social change if their own tactics, organizational structure, and interpersonal behaviour reflect, or "prefigure" the kind of society they want to bring about'. Through experimenting with alternative decision-making processes and pay schemes, CAT aimed to prefigure an organisational structure that is more consistent with the perceived values of sustainability.

The organisational structures and management systems at CAT have inevitably undergone considerable changes over the years. However, these organisational changes can roughly be divided into three broad periods: pioneering, egalitarian and managerial. These changes have been responses not only to

DOI: 10.4324/9781003207702-4

internal dynamics, but also to the external context. During the very earliest pioneering period, decision making was fairly *ad hoc*. This period only lasted for about a year. On his appointment as first director in 1975, Roderick James decided that there needed to be some formal structures. Nonetheless, as far as possible, the organisational structures were intended to foster consensus decision making. Leach suggests that 'consensus affords those affected by a decision the right to participate in making it and to block any decision they see as harmful or immoral'. Furthermore 'consensus builds strong bonds of mutual trust and solidarity' (Leach 2016: 36). In other words, consensus reinforces the sense of bonding capital and community discussed in Chapter 2.

This egalitarian phase was the longest period of CAT's history and lasted until 2010, when a new more hierarchical system of management was introduced. During the egalitarian phase, CAT experimented with consensus decision making and a relatively flat pay structure. This organisational structure can be considered as a form of communitarianism. Ralph Levinson et al suggest:

> Communitarianism maintains that the attitude of uncontrolled free-market economy and consumerism should be reconsidered. Treating the environment as an inexhaustible source of profit without thinking about the (near) future seriously endangers the planet and threatens to drive society into economic and demographic catastrophe.
>
> (Levinson et al 2020: 18)

Furthermore, communitarians suggest that the environmental, social and economic threat posed by capitalism and consumerism has to be addressed not only at global and state levels, but also by individuals and communities. Through communitarian experiments, CAT aspired to create a context that would foster green citizens who would have 'an active and committed engagement to pursuing a certain way of life consistent with a more sustainable society' (Levinson et al 2020: 20).

Until the managerial changes in 2010, there were a number of tensions between the legal structure and the way in which the organisation actually operated on a day-to-day basis. The main tension was how to balance the formal structure of CAT, as a legally registered charity, with the more cooperative ethos. Formally constituted charities are legally required to have boards of trustees who have ultimate power and responsibility for the organisation. This distinguishes CAT from a worker cooperative. In formally constituted cooperatives:

> [Workers] own their organization as they invest both capital and labour. Formally, they control the organization democratically, usually by voting the major strategic orientations and electing the managers at general assemblies, following the principle of one worker, one vote.
>
> (Dufays et al 2020: 6)

CAT workers invested their labour, but not capital. So formally they were employees of the CAT Charity Ltd, yet at the same time people at CAT did have a real sense of ownership. While workers did have considerable control over the strategic goals of CAT, this was mainly because of the goodwill of the Board of Trustees and was not legally binding. In addition, however much one attempts to put structures in place to ensure equality, some people are much better at exploiting the system than others and inevitably some unintended hierarchies emerge. Egalitarianism is also contingent on reciprocity. Responsibility to the whole must be integral to the individual's right to participate in the strategic decision making. For the most part, individuals were committed to the overall mission of CAT and there was a strong sense of mutual reciprocity, which as Robert Putnam indicates 'can facilitate cooperation for mutual benefit' (Putnam 2000: 21).

There were two aspects of the complex CAT organisation that militated against a generalised reciprocity. First, there was a constant flow of people through CAT, particularly in the very early pioneering days. The core of long-term participants had an allegiance to the overall mission and were therefore committed to the ethos of reciprocity. However, there were some individuals who turned up on the doorstep with very strong personal agendas who still had the right to participate in the decision-making process but were more committed to their own vision rather than that of the organisation. Second, when CAT got bigger and was divided into specialised departments, there was a risk that there would be a greater commitment to the department rather than CAT as a whole.

The Pioneering Days: Anarchy or Autocracy?

CAT was from its foundation a registered charity. This meant that it had to file a memorandum of association. The signatories on the first memorandum of association were Morgan-Grenville and four of his acquaintances. The signatories have legal power and responsibility over the direction of the charity within the parameters of the memorandum. All that is legally required is that the board meet annually and the charity remains financially viable. For most of its history, the trustees were not particularly involved with any aspect of CAT and tended to remain in the background. The exception to this was Morgan-Grenville as he had much more of an investment, both financial and emotional, in the embryonic project. In theory, Morgan-Grenville could have been quite autocratic. Peter Harper suggests:

> The Founder was in principle all-powerful and could have run everything as he saw fit, by *fiat*. Yet he had also bought into the alternative rejection of concentrated power and wealth.
>
> (Harper, no date)

Morgan-Grenville did not live locally but was a regular visitor in the early days, and by all accounts was actively and physically involved in transforming the site.

He clearly personally struggled with some aspects of the project, particularly the slow pace that the project was progressing and the rather scruffy look of the place. There was also some resentment of his privileged background. Mark Mathews, who arrived with his wife Mary in March 1974, recalls that:

> The place was very 'primitive' and Gerard would come up from his nice place in his smart car and do a bit of slumming with us ... In my mind there was a dissonance between those of us who were really slaving away for nothing, putting our heart into things and somebody like Gerard living in his big palace with his smart BMW. The conditions we were living in were as low as we could be and he was lording it over us.
>
> (Mark Mathews, Oral History)

Roderick James, the first director recounts:

> Gerard Morgan-Grenville had tremendous vision, tremendous enthusiasm, and there was a big gap between that and the sort of people who were available to do it ... He had a level of expectation, which actually very often could not be matched by anyone else.
>
> (Roderick James, Oral History)

James recalls that Morgan-Grenville was perceived by some of the early members as 'an autocratic controller' and that there was ultimately a showdown about who had control of CAT. He suggests that there was some resentment to Morgan-Grenville's occasional visits and his missives sent on embossed paper indicating that the standards at the Quarry were not what he expected. Very graciously after this confrontation, although still formally the chair of the board of trustees, Morgan-Grenville stood back and more or less allowed the project to progress on its own terms.

Tony Williams[1] who was the first *de facto* site manager kept a diary, which starts on 14 February 1974. There are several issues apparent in this record. The living conditions were very harsh and so the priority was to renovate the quarry workers' cottages. Clearing the site, making the cottages habitable, sorting out drainage and restoring the steep access road involved hard physical labour. There was no time for any formal meetings as such, so decision making was rather *ad hoc*. There were clearly some differences of opinion about the work, such as the need to remove trees. Williams was clearly frustrated by several aspects of the project and left in May and Gerard asked Mark and Mary Mathews to take over the site management. They managed to hold some formal meetings and to formulate some basic ground rules. For example, Mary Mathews, who took up the daily diary, records that someone spent £40 on building equipment without clearing it. As a consequence, the community held a meeting to discuss finance, and it was decided that no one could spend more than £2 without the approval of everyone. This was possible, because at the time the community was only eight strong. By the summer of 1974, Mary Mathews records that 'numbers have now reached 16 people'.[2] The central group of Mark and Mary Mathews, Gerard Morgan-Grenville and Diana Brass in this first year clearly had more of a say over the strategic direction of the nascent project.

The other thing that is clear from the diary is that there was a great deal of interest in the project from the beginning. Numerous people seemed to show up from the purely curious to ardent seekers of alternatives. Some stayed for a few days and worked hard, some had useful skills and others not and some were simply overwhelmed and fled. This constant flux of people with various levels of skills and sometimes arriving with their own agendas was recorded in the diary as a mixed blessing. Some contributed greatly to the project, whilst other visitors were regarded as a distraction. While consensus was the overriding ethos of the group, with the constant flux of people:

> What emerged was the consensus of all who happened to be there *that night*. A few days later it might be a different group of people, so 'policy' could lurch drunkenly from one extreme to another.
>
> (Harper, no date)

Egalitarian Experiments and Structures

When Mark and Mary Mathews left, it was clear that the project needed more stability and structure. As Jo Freeman in her seminal paper *The Tyranny of Structurelessness*, first given as a talk in 1970 and formally published in 1972, states: 'for everyone to be involved in a given group and to participate in its activities the structure must be explicit, not implicit' (Freeman 2021). Consequently, Morgan-Grenville invited Roderick James to be CAT's first director. The appointment of James was an autocratic decision, and so he was in many ways imposed on the community. In addition, having a director seems to imply a more authoritarian and hierarchical style of management. However, under James's stewardship, the project did become more coherent and focused.

Very early on in his tenure as director, James went on a trip on Morgan-Grenville's barge. It was their discussion on this barge trip that 'laid the blueprint for the Quarry' as 'a visitor centre showing and demonstrating [alternative technology]'. When James returned to the Quarry to inform the others about the proposal, he was met with some resistance. People were more incensed by the way the idea was imposed by a cabal and that there had been no discussion than with the proposition itself.

> There was a reaction, probably quite a pronounced reaction to the sort of autocracy, the way it was being introduced. The idea was that we should have all sat around to talk about it.
>
> (Roderick James, Oral History)

However, the blueprint was supported by the core group and the three intertwined strands of the CAT project were established – experimentation, community and education. James suggests that from then on CAT 'began to move in a much more focused direction'.

The very fact that Roderick James was given the title director by the Board of Trustees suggests that he had some authority.

> Rod was the project director, and just by virtue of having that name, there was some sense that he was leading it.
>
> (Jill Whitehead, Oral History)

James himself suggested that he was quite autocratic in the beginning. Although by all accounts he was a powerful personality, James also had the ability to foster a sense that everybody was involved and included in the decisions of the early community.

> The funny thing was that he did not come across as a leader. Although I think in some subtle way, he did lead … he gave people the impression that it was we were all in it together.
>
> (Jill Whitehead, Oral History)

James instituted two important interconnected initiatives that laid the foundation of a consensual decision-making process. The first, suggested by Morgan-Grenville, was that there should be formal weekly meetings. The second was a system of rotas. Both these initiatives cultivated senses of egalitarianism, sharing and belonging.

Interminable Meetings: Consensus or Compromise?

James established weekly meetings on Mondays. These weekly meetings were regarded almost unanimously as not only successful but also incredibly time consuming. One of the major frustrations was that the meetings could be hijacked by individuals, even those who were just passing through.

> Quite often we allowed short term volunteers, vociferous short-term volunteers who were quite eloquent to come in. And actually they could knock us off course because you could spend a lot of time arguing with people who had no responsibility or involvement really with [CAT] … We didn't have an authority or structure that was strong enough to actually withstand some of these people who blew in and would also start convincing other people.
>
> (Roderick James, Oral History)

Nonetheless, weekly meetings entailed that people felt involved in the decision making and contributed to creating strong bonding capital.

James recollects that he spent a great deal of energy prior to any meeting trying to actively persuade people of his point of view.

> I spent every evening lobbying people because when you had the meeting you had to have people on board with what you were doing. You had to talk to everyone because you wanted everyone to be part of it. But, you know, as the director it was a very time-consuming business.
>
> (Roderick James, Oral History)

The other aspect of the meetings was that when James became the director, although there was still some presumption that everyone was expected to be able to do a bit of everything, there was the beginning of specialisation. For example, no one had the expertise that Bob Todd had in electrical engineering. Yet individuals could express opinions on topics in which they had no expertise. Overall, there does seem to have been a respect for people with a specific specialism.

> People were respected within their own disciplines. But I think at the same time, anyone could volunteer opinions or input ideas on everything, really.
>
> (Roderick James, Oral History)

It is not really the final decision itself that is significant in consensus decision making, but the sense of participation. Serge Moscovici and Willem Doise observe that 'what effects consensus and makes it convincing is not the agreement itself, but participation by those who arrived at it' (Moscovici and Doise 1994: 2). This sense of having, or at least the possibility to have, an active input into the decision making created a sense of egalitarianism and participation. While there were obviously many differences of opinion, there was an overall ethos of 'we are all in this together'. Involvement, egalitarianism and participation also entail shared responsibility. There are two aspects to this responsibility that are best identified when things go wrong. First, if things do go wrong, then the onus is on the whole group, irrespective of whether the misfortune was the result of poor decision making or unavoidable external circumstances. For example, when there was a cash flow crisis in the 1990s, everybody took responsibility. All those who could afford agreed to to have their salaries suspended.

> What I loved about CAT and the whole co-operative working thing was that you were sharing responsibility. There was a whole group of other people, so it was never a question of the buck stopping with me as an individual.
>
> (Roger Kelly, Oral History)

Second, if the group seems to be on the verge of making a poor collective decision, then any individual who recognises the group decision as a miscalculation has a responsibility to point this out at the meeting. However, it is often difficult to identify if an individual is simply being obdurate or has got a genuine concern. Normally, if an individual or a very small minority found themselves at odds with the entire group, they tended to back down to ensure full consensus. This demonstrates that loyalty to the group more often outweighed personal convictions. This is not to say that individuality was subsumed within a sort of hive mind. Individuals were always given the opportunity to express their opinions, often vociferously and at length. Debate is an absolute prerequisite for consensus. Even if some individuals are not necessarily in full agreement with the final decision, they have a sense of participation and they have willingly consented to the outcome. There was a sort of unwritten agreement that if an individual opposed something that the rest of the group supported:

> You would stand aside and say, okay, I do not like this, but I won't block the consensus.
>
> (Peter Harper, Oral History)

This is more than just acquiescence. Moscovici and Doise state that 'consensus goes well beyond mere acceptance and agreement … The convergence of individuals, which binds them mutually as regards interests and ideas, fosters their confidence in one another' (Moscovici and Doise 1994: 4).

Consensus decision making assumes that the more minds that are actively involved in decision making the more likely that a good decision will be made. However, that is not necessarily the case. Collective decision can tend towards compromise. If the group is divided about a particular decision, the ideal is that discussion will lead to both sides of the debate recognising the value of the other's point of view and subsequently agreeing to modify their position. 'Each individual acknowledges a certain merit in the position of others without being forced to repudiate his own' (Moscovici and Doise 1994: 8). However, the happy medium is not always the best decision. 'It is precisely through compromise that the group can allegedly manage individuals and make them fit in to a common reality' (Moscovici and Doise 1994: 30).

> It was difficult to make plans because the structure of a cooperative tends to be one of universal compromise. So the organization sort of sits in the middle of a number of disparate passions and enthusiasms and projects.
>
> (John, interview February 2021)

Roger Kelly, the third Director, indicated that consensus was very important for him, and that he was amazed how well it actually worked. Kelly made strategic use of the general predilection for people to tend towards compromise.

> I would present three of what you could sort of crudely describe as two extreme options with a compromise in the middle, knowing that the chances are that is the compromise in the middle.
>
> (Roger Kelly, Oral History)

There was one particular time that has become part of CAT mythology, when consensus failed to be achieved because an individual stood their ground in opposition to the group. While the actual details of the decision, which was something to do with visitor parking, seem to have been largely forgotten, the fact that an individual was not swayed by the group is indicative of how significant it was. Jeremy Light, who coordinated the biological aspects of CAT between 1976 and 1973, recalled that he firmly believed that the group was making a poor decision:

> I was absolutely certain that I was right, but it took quite a lot of courage to stand for that ... I was the one who opposed consensus. So suddenly the onus was on me to prove my point. It is essential for a consensus system to work that people realise, if they are the only objector there is a great responsibility on them to sort that out quickly. You have to be very careful not just to be obstinate.
>
> (Jeremy Light, Oral History)

Everyone was really shocked by Jeremy Light's apparent intransigence. Everyone agreed to delay a decision, but if no agreement was reached by the next meeting, the issue would be voted on in a secret ballot. However, it seems that Jeremy Light was correct in his objection.

> We began to realise that Jeremy had understood something that nobody else had.
>
> (Peter Harper, Oral History)

If there had been no individual as single-minded and courageous as Light, the group could have easily made a very poor decision.

Another aspect of consensus decision making is that through group discussions undue significance can be given to things that are quite trivial. Consequently, a great deal of time was often taken up discussing things that were not particularly important. The more time spent discussing relatively inconsequential issues, the more significant they appear to be. Several people recall spending almost a whole afternoon discussing where to place a spice rack in the communal kitchen in Tea Chest. There are three issues about consensus decision making: are there any criteria for determining what is trivial and what is significant; what decisions have consequences for the group as a whole and what impacts only an individual or a subgroup; and who has the authority to make a decision.

> I think a very important part of consensus is that you reserve that principle of consensus for the really important things. You don't have to use it for day-to-day decisions which are much better left to individuals or small groups that you trust to make decisions by themselves.
>
> (Roger Kelly, Oral History)

It is of course unclear where the divide between 'really important things' and 'day-to-day decisions' lies.

Group decision making can be very time consuming and inefficient. This was noted in many of the oral history interviews. However, sometimes a decision that might have major consequences for the entire group has to be made quickly, and this raises a further dilemma. Occasionally, an individual or small subgroup perceives that a significant course of action would be beneficial to the group. However, often because of pressure from the outside, some decisions have to be made quickly and there is no time to go through the consensus decision-making process. The individual or the small group then must determine whether to make an executive decision or consult the whole group and risk losing that benefit. This is, as Leach (2016: 39) suggests, a choice between 'the purity of means and the primacy of ends'. Purity of means is often utopian and primacy of ends tends towards pragmatism – and it is always a dilemma of where the best balance lies. Purity of means can entail failure to actually deliver the agenda of sustainability, but the primacy of ends can entail too great a compromise. Furthermore, different people have different perspectives. For some, the purity of means was absolutely sacrosanct, as they considered that consensus decision making was not only integral to CAT's ethos but also intrinsic to sustainability. For others, executive decisions had sometimes to override the precedence of process, not only for the benefit of CAT as an organisation, but also because it was perceived as advancing CAT's sustainability agenda.

Group decision making was often successful when it came to short-term issues but was not so effective when it came to major long-term strategic plans. Not everyone is good at long-term strategic thinking. The immediate often takes precedence over the long-term. There also tends to be a degree of reticence in making a final decision to invest in a long-term project that has the potential to damage or even destroy the viability of the organisation. Group decision making can often be quite risk averse. Recalling the debates over the WISE building (discussed below), which was a major project proposal and necessitated CAT having to borrow substantial funds. Phil Horton who was the project manager observed:

> We always used to get to a point in a debate where somebody would say, 'so are we pressing the go button?'.
>
> (Phil Horton, Oral History)

However, the project was eventually agreed by consensus despite some people being uncomfortable with the decision.

While the Monday meeting continued, it was primarily focused on immediate practical tasks, particularly ensuring that the displays were working, and that the site was tidy. Consequently, monthly permanent staff meetings (PSM) were established to discuss long-term strategies, finance and so on. John Urry suggests that while the PSM could be tedious and, after being at CAT for a while, it was possible to predict what everybody would have to say, nonetheless:

From a psychological point of view it gave the feeling that everybody was a stakeholder, in what we were doing, it kept everybody involved, it kept everybody feeling that they mattered, that they had some contribution to make.

(John Urry, Oral History)

Ogres and Rotas

There were numerous rotas. Most of the rotas involved basic chores that it was felt needed sharing. Peter Harper (2016) lists five rotas that were maintained between 1974 and 1992: Staffing reception, cleaning the communal dining room, cooking lunch, baking bread and ogre duty. Rotas are intended to inculcate a sense of sharing responsibility and belonging as well as ensuring that basic chores are completed. Keeping the site clean, tidy and well maintained is particularly important as CAT is a visitor centre. The main aim of the visitor centre is to try and convince the public that human activity is harming the environment and that there are solutions to the damage caused by human actions to the planet. These goals were not going to be readily achieved if the site looked a mess and if displays were not working properly. Roderick James instituted a system that he called the Ogre. This was derived from the New Alchemy Institute, one of the main inspirations for CAT. The Ogre system was linked to the weekly Monday meeting, which gradually became more focused on immediate work tasks rather than long-term strategic direction. Everyone was expected to take a turn in being the weekly Ogre. The designated Ogre of the week walked around the site on a Monday morning to assess what needed tidying and repairing. The Ogre would then report to the work meeting what work had not been completed from the previous week, inform the group of what new jobs required attention and identify who was responsible. These jobs were recorded and, this being before computers, the lists of work were kept on clipboards in the main office where everyone could see them. In theory, the Ogre was conferred with an absolute authority for the week. The term itself is intended to demonstrate that the role of the Ogre would be imbued with certain ferocity to ensure that mundane chores were completed.

This system worked well up to a point and remained in place for most of the egalitarian phase. However, there were two issues with the system. First, while the group delegated authority to the Ogre, they did not have any real power. According to many accounts, certain jobs remained on the list for months without getting done. Second, different Ogres had different standards and were not necessarily sufficiently authoritarian.

The Ogre had absolute authority … Of course, different people had different standards. The problem is that on the whole people living at the Quarry got used to a standard which was completely different from society … Everyone got the chance to be the ogre, but some people were not prepared to be as tough as they needed to be.

(Roderick James, Oral History)

Jill Whitehead, in her oral history, observed that she thought that introducing the Ogre system was a very shrewd move by James as 'it took it away from him always having to be the ogre'. In other words, by delegating responsibility and authority to the staff body as a whole, it mitigated the sense of having an autocratic boss. This gave people a sense of having much more control and involvement in the overall project.

One of the Ogre's duties was to ensure that the public toilets were cleaned. In other words, everybody was expected to not only assume the mantle of authority, but also perform what is commonly regarded as the most menial of chores. Roger Kelly noted:

> I actually spent quite a lot of time cleaning the public toilets at CAT. I actually liked doing it, not because it's a particularly pleasant job, but because I really believed that everybody should have their share of doing it. And I didn't want to be in a position where I was exempt from doing that and other people did simply because they had a different job title.
>
> (Roger Kelly, Oral History)

The Ogre and toilet cleaning were not the only rotas. The other important rota was preparing the lunchtime meal for everybody. Taking turns in cooking and eating together was, as I indicated in Chapter 2, an important way of creating bonding social capital. This combination of rotating both authority and basic jobs reinforced both the egalitarian and sharing nature of the project. The rotas, and in particular the Ogre rota, 'helped maintain the general feeling that "we are all the management; we are all the workers"' (Harper no date). In other words, there was a sense that CAT was a workers' cooperative although it was not legally one.

Streamlining Consensus

As the group grew in numbers and the weekly meetings became increasingly focused on the work needed to maintain the site, it was decided that there needed to be some other mechanism for decision making. The major challenge in the early days was mainly about what we would now call human resource issues, so Roderick James instituted a small personnel group. Pete Raine, who became the second director, supplemented the personnel group with an elected management group, which got called Overview. All members of the permanent staff could stand for election, and indeed there was a sort of expectancy that everyone would at some point put themselves forward for election. There were four members of the group. Each member would be elected by a secret ballot for 18 months and then they would have to stand down, although they could stand for re-election. The director was the permanent chair, which gave this management group a degree of continuity. As Peter Harper observes, with the formation of Overview 'it was noticed how much more efficient it was to delegate many decisions to a small group, rather than spend a day a week thrashing things out *en masse*' (Harper no date).

Roger Kelly observed that not everybody wanted the responsibility of being on Overview. Kelly recalls a study, which he attributes to Braziers Park, an intentional community and educational centre established in the 1950s, which suggests that only five percent of people really want to take on any sort of leadership role. This is clearly a challenge for instituting an elected management system, as not everybody necessarily wished to take on the responsibility of leadership. While some individuals were reluctant to take responsibility, there were also some individuals who clearly relished leadership roles. Therefore, inevitably, while the pool of people for Overview included all the permanent staff membership, some individuals tended to be frequently re-elected, whilst others rarely stood, if at all.

As there seemed to be just a core group that volunteered to serve on Overview, it was thought that it did not necessarily represent the whole staff body. Consequently, Overview was changed to a system in which CAT staff were divided into four distinct groups, each of which had a representative that served on Overview. Peter Harper views this as a good idea, but fundamentally flawed.

> As 'delegates', the new members felt that they should press the interests of their own 'constituencies', rather than acting in the interests of the entire organisation. [This] showed that what might have been thought of as core values were not universally or automatically shared and the system returned to elections from the entire staff body.
>
> (Harper, no date: 18)

It was not always clear which decisions could be devolved to Overview, and which should be decided by the group as a whole.

> There were too many small to a full staff meeting where the Overview group should have made the decisions themselves.
>
> (Phil Horton, Oral History)

The principle developed that there were really three tiers of decision making: day-to day decisions should be decided by individuals or the small groups that were involved; larger and more complex decisions should be resolved by Overview; and long-term strategic decisions or if there were any objections to Overview decisions were debated by the entire staff body. However, it was not always clear what criteria determined what should be decided in what arena.

Issues that seemed to be of no consequence to one member of staff, who then makes a unilateral decision, may be seen as highly significant to another individual who is then annoyed by not being consulted. For example, there was an occasion when Clive Newman, one of the site engineers, demolished a small building that housed a compost toilet that several people were particularly fond of. The demolition was done as part of the revamping of the site for Gearchange. This raised the issue of disciplinary action. Newman recalls that 'no one had ever broken the rules before because it was inconceivable'. He admits that he was in a hurry to complete a job, and that there was sometimes a clash between the pragmatism of getting a job done quickly and the principle of consensus, which sometimes he suggested was like 'swimming through treacle'. The issue was not so much the actual demolition of the building, but that he had circumvented the hallowed principle of allowing everyone to have an input into decisions. Newman suggests that 'I was responsible for CAT introducing a disciplinary procedure' (Clive Newman, Oral History).

As CAT developed, the staff size grew and the nature of the organisation became more complex. To address the first issue, Overview decisions were expected to be ratified by the full staff meetings. However, even this could be time consuming. Consequently, a system was introduced where the minutes of the Overview meetings were posted in the central office where everybody had access. If anyone opposed any of the decisions made by Overview, they wrote an objection next to the particular issue. This was then taken to the staff meeting to be discussed. If no one objected, the decision of the Overview was deemed to be ratified. This also pre-empted the Overview group from making executive decisions behind the larger group's back. Peter Harper calls this system 'passive ratification'. He suggests that 'the key here is that objecting to decisions of elected groups must be made *somewhat* difficult, but not impossible'. He observes that once this system was introduced 'nearly all Overview decisions sailed through *nem con*' (Harper no date). This also entailed that once a decision had been passed no one would have cause to object that they had not been consulted.

Everyone Is Equal, But Some Are More Equal Than Others!

While every effort was made to ensure that CAT was as egalitarian as possible, absolute equality is a utopian pipe dream. There were both formal and informal hierarchies. CAT, as a charity, had to have a board of trustees who have legal responsibility for and authority over the organisation. However, for most of its existence, the trustees took a back seat and, other than the legal requirement for an annual meeting and ensuring that the charity was financially viable, pretty much left CAT to run on its own. Although Morgan-Grenville was viewed as being fairly autocratic by some and clearly had some authority over the direction of the fledging organisation, he did step back from controlling the organisation when challenged. In the very early pioneering days, decision making largely involved everyone sitting around a table discussing the issues. However, it was clear that the stable core group had more of a say than the more short-term volunteers.

Formally, Roderick James and his successors were directors, which implied a degree of authority. Each of the directors had their own management style and strategies. James, as indicated above, sought consensus through lobbying. His successor Pete Raine instituted the Overview group, to try and streamline the decision-making process. There was a gap between Pete Raine and Roger

Kelly, the third director. Kelly, in his oral history, notes that CAT successfully continued for 18 months without having a formal director. Consequently, he reflected that the role of the director was to take a step back from the day-to-day running of the organisation and be focused on the more long-term strategic direction. Kelly described his approach to the role as leading from behind in order 'to get the best out of everybody else'. The oral interviews of people who experienced more than one of the first three directors suggested that Kelly's approach was the most democratic. Overall, the idea was to find some sort of balance between the organisation being a rudderless ship and a more formal managerial system. Each of the first three directors adopted slightly different strategies and approaches when trying to find this difficult balance.

In addition, there was inevitably a distinction between the influence of volunteers and casual seasonal workers, and the permanent staff members. Peter Raine, the second director, instituted a quarterly PSM. He suggested that these quarterly meetings were 'the most important meetings by a country mile'. These 'were to review progress and to review budgets and see where we were financially and what our plans were for the future'. These quarterly meetings were only open to permanent staff, and Raine indicates that occasionally volunteers or the casual staff complained that they were excluded (Pete Raine, Oral History).

While the aspiration was to be as egalitarian as possible, inevitably informal hierarchies emerged and some people had a greater voice in the decision-making processes than others. There were three types of informal and often unacknowledged sources of power at CAT – the authority of expertise, the authority of longevity and the authority of self-confidence. For example, there was a tacit acknowledgement in the 1980s and 1990s that 'engineers were king'. This was in part because site maintenance and the ability to maintain and repair the technology necessitated the expertise of engineers. It was, for example, difficult to challenge an engineer on the supply and availability of power on site. Sometimes, this also entailed that more leeway was given to the engineers about missing some of their rota duties, as it was considered that their expertise, such as the ability to fix the site power, often had to take precedence over the more mundane quotidian chores.

Longevity could also be regarded as a source of authority. The longer an individual had served on the staff, the more credence was sometimes given to their input into the discussion. While this could be beneficial, as it often avoided replicating things that had been tried in the past but not worked, it could also engender a conservatism that could stifle new innovative voices.

> I would sit there in meetings with the old timers, who would all be sitting there going, 'oh yeah, we know what we are doing, you know' and just making my life difficult.
>
> (Rachel Lilley, Oral History)

No matter how egalitarian a structure is, certain individuals can use the systems more effectively than others. Certain people are more confident and eloquent while other individuals are more apprehensive about speaking in meetings. It was suggested in several of the oral history interviews that the most confident individuals could push through decisions. Lynda Wenman, who ran the Quarry Café for many years, noted that meetings were 'challenging for someone who was quite shy and maybe that wasn't really recognised enough' (cited in Shepherd 2015: 75). In addition, some people's roles and responsibilities were more flexible than those of others. Those involved in public facing roles, such as working in the restaurant or bookshop, were not always able to attend meetings.

The consensus decision-making process:

> Worked better for some people than for others. While it gave everybody the opportunity to express their point of view … It was those people who were most willing to speak the most who tended to get the decision through.
>
> (Phil Horton, Oral History)

Pay

The other really important aspect that fostered senses of egalitarianism, sharing and participation was CAT's experiment with pay. In the very beginning, the pioneers were paid almost nothing. However, it became apparent that to maintain a stable core, people had to be paid some sort of a stipend. Initially, Roderick James and Bob Todd proposed that everybody should be remunerated sufficiently to cover the basic essentials to live. Basic needs were adjusted for individuals who lived off-site and/or if they had children to support. Above the basic salary, there was a very small differential for what individuals brought to the project. However, when Todd and James presented their figures, people objected as they felt excluded from the decision. So Todd and James drew up a list of ten characteristics that ranged from whether an individual had a particular skill to whether they worked particularly hard – and the group assessed everybody including themselves according to these criteria. James notes that although individual assessments were predominantly consistent not only with each other but also with the original figure, he and Todd first mooted:

> It was interesting that one person actually valued herself at the bottom and everyone else valued her as being extremely good. And one other person valued themselves as being extremely good and were valued by everyone else as being extremely poor.
>
> (Roderick James, Oral History)

Eventually, this system was abandoned as being too complex, and when I worked at CAT in the mid-1980s, all permanent members of staff received the same salary with a mortgage allowance for those living off-site and extra for up to two children. However, if you were single and lived on-site, as I did, you had a little more disposable income.

The basic principle of the pay was each according to needs, and it was assumed that if you had children your needs were greater. However, in the 1990s, there was a discussion of whether the child allowance should continue. Some of the individuals without children argued that they also had needs that were not catered for, and that it was fundamentally unfair. This was such a fundamental challenge to the received wisdom that it was decided that it should be put to the vote. This ended with pay parity, and members of the permanent staff were effectively paid the same, regardless. This almost flat pay structure lasted for a number of years,[3] until the introduction of the Associate Co-operative Members (ACM) as opposed to Full Co-operative Members (FCM) who were paid slightly more as they were expected to take greater responsibility. In other words, the underlying principle started to shift from 'each according to their needs' to 'equal pay, equal responsibility'.

Egalitarianism and this relatively flat pay structure was to a certain extent difficult to maintain, as it is embedded in a social system that values certain tasks differently. In the wider social context, a brain surgeon is valued more than a cleaner, and a CEO is valued more than a shop floor worker. These values are reflected in the vast discrepancies between the higher and the lower paid in society. Roger Kelly recounted the experience of an equal-pay cooperative that was primarily composed of scientists with PhDs. In the initial years of this cooperative, the group did their own cleaning. However, they decided that this was taking too much of their valuable time. The group decided to hire a cleaner but were still committed to the idea of pay parity, and consequently they offered to pay 'about five or six times the going rate for cleaners and they were inundated with applications' (Roger Kelly, Oral History).

At CAT, the gardeners and catering staff were paid approximately the market rate, but the salary of the highly skilled engineers, architects and builders was considerably lower than they could have earned elsewhere. To a certain extent, this was compensated for by giving highly qualified and experienced members of staff more leeway on missing core rota responsibilities, and more credence was often given to their opinions in meetings. Nonetheless, there is a common refrain in the oral histories that no one worked at CAT to get rich. In addition, as Rick Dance, who was an administrator at CAT for many years, observed, 'I knew no one was getting rich at my expense' (Rick Dance, Oral History).

Pay parity is not only about who is valued, but it is also about fairness. If resources are limited, then they should be shared fairly. In many ways, sustainability might be considered to be underpinned not only by the fear of environmental catastrophe, but also by a sense of fairness. If resources are limited, then these resources must be shared fairly. CAT's experiments with consensus, egalitarianism and a relatively flat pay structure, although inevitably flawed, were intended as a model for what a sustainable society might look like. CAT's experiments might be considered an aspect of utopianism that Wright terms emancipatory, 'a central moral purpose in the production of knowledge – the elimination of oppression and the creation of the conditions for human flourishing' (Wright 2010: 10).

The Demise of Egalitarian Experiments

Around 2010, CAT faced a tsunami of factors that challenged the egalitarian ethos and rendered it unsustainable. Some of these factors were internal and some were external; some were of CAT's own making and some were beyond CAT's control. Before continuing with this section, it is important to stress that this was the most painful period of CAT's history and was incredibly distressing for many of those caught up in the events. Internally, CAT had expanded and had reached nearly 150 full-time staff. Concurrently, CAT experienced a major financial crisis, which was much more severe than any other previous crisis. This crisis necessitated, for the first time since Morgan-Grenville had stepped back from active involvement, the intervention of the Board of Trustees. It also necessitated redundancies and some job restructuring. The external context had also dramatically changed. By 2010, there had been a sea change in the attitudes on environmental issues and CAT had to determine how best to respond to these changes.

These contextual changes were economic, technological, political and social. By 2010, the green economy had begun to emerge. In the late 20th century, there were very few employment opportunities for engineers and so on who had an interest in renewable energy and CAT was one of the few employers. However, by 2010, the sector had expanded exponentially and there were many more green employment opportunities. Furthermore, the recession had hit, so the combination of the fact that AT was not exotic anymore and many people had less disposable income saw a gradual decline in visitor numbers. The year 2010 also saw a decline in postgraduate students. Linked to the emergence of the green economy was the rapid advance in the technology of renewable energy. The development of wind turbines that could produce megawatts of power and photovoltaic cells being increasingly effective meant that wind and solar energy were now an economically viable alternative at a national scale. Concern about the environmental devastation of human activity was also well and truly embedded into political discourses at local, national and international levels. Finally, various environmentally friendly lifestyle choices, such as recycling and energy conservation, had become socially acceptable.

Restructuring Management

As CAT grew in size and complexity in the late 1990s and early 2000s, it became apparent that the management system was no longer effective. Not only were there more staff, but also the annual turnover had increased to approximately £4 million. Most of the established ways of doing things at CAT had been developed in the 1980s and 90s when there were far fewer staff, and there had really only been a few minor tweaks since. Many people recognised that these decision-making systems were no longer fit for purpose and a long drawn-out review of decision making took place, which resulted in major organisational changes.

Peter Harper cites the idea of Dunbar's number as a reason why CAT's consensus decision-making processes eventually failed. The anthropologist Roger Dunbar suggested that humans can only maintain 150 stable relationships. Harper conjectures that CAT's consensus decision-making process became unworkable at the point where the staff reached about 150. He suggests that 'the complexities of this scale did indeed seem to be more than could be coped with by part-time managers' (Harper no date: 22). There are successful workers' cooperatives that have considerably more than 150 members. For example, Mondragon, which is the world's biggest workers' cooperative, is actually a federation of smaller sub-cooperatives.[4] However, CAT, although divided into separate departments, is too small to establish a federation of semi-autonomous units and appeared to be too unwieldy to operate as a single cooperative.

Meetings of all the staff became increasingly challenging and drawn out as there were many more attendees. There was even an issue of finding a room that could comfortably accommodate all the permanent staff. Serving on Overview involved ever more complex managerial, HR and fiscal decisions that many staff did not have the necessary skills or training for. In addition, with a larger staff body, there was also more likely to be disagreements and even resentment of Overview decisions. Being on Overview became increasingly time consuming and people who were very passionate and dedicated to their own work areas did not necessarily wish to spend the time being a manager. Furthermore, with an elected management group, there was a turnover of individuals, and this was a challenge to ensuring continuity.

> The Overview role just didn't work. It took it far too much time it caused far too much stress for the people who were doing it … [We] did agree by consensus that we needed to have a new structure.
>
> (Phil Horton, Oral History)

Five different ideas on how to restructure the decision-making process were proposed and presented to the full staff body. The proposals ranged from small adjustments to root and branch changes. Through discussion, these were eventually whittled down to two, which were then formally voted on, and which itself was a break from CAT's usual consensus decision-making practice.

The overall feeling was CAT required people with more experience and expertise in management. A two-tier management structure was eventually approved in which the Directions Team would be elected from the general staff, and the Operations Team would be responsible for the basic management. The Directions Team would have a long-term view of strategic goals, and the Operations Team would be responsible for the day-to-day management within the parameters set by the elected Directions Team. The Operations Team would comprise four people, each of whom would be responsible for one area: site maintenance, education, marketing and fund raising, and finance and human resources. Although Phil Horton, who designed this system, acknowledges that his proposal had been much altered during the consultation process, he nevertheless suggests that:

> Effectively the Operations Team would be working for the CAT staff to make sure that the organization was running smoothly and doing what we wanted.
>
> (Phil Horton, Oral History)

This system was sometimes represented as the Direction Team being analogous to MPs and the Operation Team being the civil servants.

The proposal immediately encountered problems in implementation. First, it proved difficult to recruit experienced managers on a CAT salary, although eventually four candidates were appointed. Second, there was an ambiguous relationship between the Directions Team and the Operations Team that was never fully resolved. There was also a somewhat fraught relationship between the Operations Teams, who were recruited for their management expertise, and the CAT staff, who were accustomed to operating fairly autonomously. Joan Randle, a long-term member of CAT, observed 'we wanted managers, but we did not want to be managed'. However, the most significant problem for this new system was the severe financial crisis that threatened CAT's survival.

> I was on the Directions Team for a little while and we really struggled to get policy and strategy going because all we were dealing with was immediate problems of how are we going to find the money to keep the organization going.
>
> (Phil Horton, Oral History)

A member of the newly appointed Operations Team recollects that:

> The very first thing was a meeting at which we had to discuss whether we had to make a lot of staff redundant or reduce their hours. So, it was not quite what I had anticipated when I had dreamed all those years of coming to work for CAT.
>
> (Sally, interview May 2021)

The End of the Flat-Pay Structure

In 2001, a maverick tutor from the University of East London (UEL) began bringing students on a sustainability Master's programme to CAT for one week every month. CAT would then invoice UEL for every student. In 2007, CAT decided to form the Graduate School of the Environment (GSE), as it could then charge the students directly, rather than simply charge UEL for accommodation.

CAT would then directly employ the academic staff and the degrees would be validated by UEL.[5] The GSE brought in a considerable income for CAT, and it was able to develop more postgraduate courses (see Chapter 4). However, the salary of the academic staff who came to teach at CAT was protected under a UK Government scheme known as Transfer Undertaking: Protection of Employment (TUPE). This meant that academic staff, who were now directly employed by the GSE, were being paid considerably more than the CAT salary. Academic staff are on a nationality recognised pay structure, which has incremental increases related to experience, whereas for those working directly for CAT there was no system of yearly increases in salary. This began the unravelling of the flat pay structure and caused some resentment. It was decided that although it was part of CAT, those employed directly by the GSE would not have any input into the consensus decision-making processes, and this alienated some of the academic staff and contributed to a rift between GSE academic staff and the rest of CAT. A lecturer on the GSE programme recalls:

> We were not even allowed to go to monthly meetings. We were not part of CAT you know. We were cut off, which that was another huge tension because that felt really weird and definitely felt as though we were resented by CAT. But then we were earning a different salary scale. There were those working at the Graduate School of the Environment being paid academic rates; full co-op members being paid one rate; associate co-op members another; and then there were the casual staff. So that the principal all are being paid equally had gone completely.
>
> (Kelvin Mason, Oral History)

Conversely, a long-time member of CAT staff comments:

> It had a dramatic effect psychologically, I think, and it certainly did for me that suddenly I was up there working alongside people who were getting paid three or four times as much as me.
>
> (Rick Dance, Oral History)

Not only was the pay differential between academic staff and the rest of CAT a bone of contention as it seemed to directly challenge the egalitarian ethos of CAT, but the way in which the decision was made also caused contention. The decision to create the GSE did not go through the usual consensus making process. This decision to create the GSE had to be made quickly. CAT had to negotiate with UEL, and higher education (HE) institutions are notoriously hierarchical:

> I don't think you can negotiate with a vice chancellor. I mean, this is the other thing, isn't it, you have to have a mandate to go and talk to a university and say, we would like to take this course from you and have this kind of arrangement. But actually I have no authority to discuss this issue

because I have to go back and talk to a hundred people about it and I'll let you know in about six months if they've decided to go ahead with it.

(Joan Randle, Oral History)

The small group who were aware of the TUPE legislation and also that there would be considerable time-consuming debate about breaking from the CAT tradition of a relatively flat pay structure opted to approach the trustees directly in order to expedite the decision. Directly consulting the trustees was itself a break from CAT tradition, and particularly upset those who were more inclined to the purity of means as opposed to the primacy of ends. The trustees agreed that the development of the GSE, even though it involved having to pay academic staff considerably more than CAT wages, was a good idea and the decision was presented at a full staff meeting as a *fait accompli*.

The Saga of the Wales Institute for Sustainable Education (WISE)

From almost the very beginning, CAT had residential courses both for schools and colleges, and the general public (see Chapter 4). However, the accommodation was very basic hostel style. Furthermore, there was a shortage of good teaching spaces. Consequently, as a concomitant project to the development of the GSE, it was decided that there was a need for a purpose-built lecture theatre and upgraded accommodation. This building is commonly referred to as WISE – the Wales Institute for Sustainable Education.

A number of workshops were organised at CAT with Pat Borer, who had been a long-term member of CAT but who now had his own architectural business, and David Lea, who had worked with Borer on designing the stations for the cliff railway. Borer and Lea worked on the ideas that were suggested during these meetings and came up with the final design. However, before investing in a project that had a starting estimate of around £3 million, CAT decided that it needed to own the Quarry that they were still leasing from the Beaumonts. They were successfully able to do this, buying the site in 2004, through some very generous donations from supporters and by mortgaging the Quarry Shop in town.

There are two significant aspects of the building of WISE to consider in relation to the current discussion (see Chapter 4 for further details about the building itself). This was a very large project, and for the first time in its history, CAT decided that it could not be done in-house. However, as the proposed building technique was relatively unusual (see Chapter 4), the challenge was to find a contractor that had the ability to construct it. WISE was also a much bigger and more expensive project than any CAT had ever previously undertaken, and so for the first time in its history, CAT had to borrow money. Several tenders from building contractors were considered. Eventually, the contract was awarded to Frank Galliers, a family run business in Shrewsbury, a small market town on the border between England and Wales and which was therefore fairly local. According to Phil Horton, who had been appointed the project manager for WISE, Galliers was keen to get into eco-building, and seemed to be very easy to work with. However, before work had started, Gallier sold his family company, though CAT was still happy to sign the contract with the new owners.

Horton recalls that the project came in at about £4.5 million. CAT had raised some money through soliciting donations and a European funding grant, but had to take out a mortgage for £1 million. This was a break from CAT tradition as CAT had never borrowed money in the past for a project. Work started on this massive project in 2006. However, almost immediately there were problems between CAT and Galliers. According to Horton, the new owners of Galliers attempted to charge considerably more for various things such as timber than was in the original budget and sent a site manager who had no experience of the rammed earth building technique. Galliers walked off site on several occasions claiming that CAT owed them money. CAT issued Galliers with notices to comply with the contract and they would return to the site before CAT could take legal action. The work slowed down as Galliers sent fewer workers to the site. However, CAT was under pressure to complete the project because of the European funding deadlines. This eventually came to a head.

> We issued Galliers with notices because they were not building to their agreed program. There were lots of faults in what they had done, and they were not responding to any of our correspondence about it. So, we issued them with a notice to quit. It was quite intimidating, really, because we went and locked up the site.
>
> (Phil Horton, Oral History)

CAT took Galliers to court and won their claim for £450,000. However, Galliers then went into administration never paid, leaving CAT out of pocket and the WISE building unfinished. CAT was able subsequently to raise the money, mostly through very generous donations, and turned to C. Sneade Ltd who had come second in the original tender bid. Sneade not only had to complete the building, but also had to rectify the errors made by Galliers. By all accounts, Sneade were much more conscientious and better to work with.

At around the same time, it was discovered that CAT's finances were considerably worse than anyone had imagined. This was in part because of the unfortunate saga of WISE, but also in part because of a decline in student numbers. These problems were also compounded by a basic accounting error. No one at CAT, at this time, seemed to have had the necessary expertise to manage the increasingly complex finances of an organisation that had grown to such a size and complexity. CAT's financial difficulties were further exacerbated by the global economic crisis. Although CAT had been in serious financial difficulties before, the fact that CAT had borrowed a considerable amount from banks to fund WISE made this crisis much deeper and had the potential to bankrupt

CAT. The staff once again elected to take a voluntary cut in salaries, but this time the sacrifice was not sufficient.

The Board of Trustees was forced to take a more active role than they had at any time in CAT's history. One of the trustees commented:

> I don't think either the trustees or indeed the staff had realized that once you are in hock with a bank life is a bit different.
>
> (Mick, interview May 2021)

These crises led to almost the entire board of trustees resigning, as they felt that they did not have the prerequisite skills to deal with this unprecedented situation. Following the appointment of a new chair and board, the trustees hired an accountant to investigate CATs finances, which revealed that there was a much more severe crisis than anyone had envisaged. There seemed to be only two options: to place CAT into administration or to try and extend the credit from the banks. The trustees approached the banks who agreed only if there were some root and branch changes to the way that CAT was organised. The banks were concerned about their investment and demanded a clearer line of management and accountability, if CAT was to avoid being taken into administration. The consultant accountant was appointed as a CEO, on a salary far higher than the CAT salary, a more conventional hierarchical management structure was instituted and the organisation was slimmed down. Departments were closed, including the Quarry Shop and café; some staff were made redundant or placed on zero hours contracts; and the site community was disbanded. This seemed like the end of CAT as an alternative organisation. It was a dark era for many, not least because they and/or their friends had lost their jobs, but because CAT had had to sacrifice consensus decision making, a relative flat pay structure and the egalitarian ethos that many thought was integral to CAT's sustainability agenda. Peter Harper (no date) observed that this looked like 'unconditional surrender' and it seemed as if 'the organisation had turned into the opposite of what it set out to achieve thirty years earlier'.

Despite the undoubted distress, anguish and sense of betrayal to CAT's core values, the organisation has survived, albeit in a very different form. As Harper notes, CAT 'has continued to function and generate what are arguably its most important and distinctive products: residential courses for the general public, postgraduate courses in the GSE and the *Zero Carbon Britain Reports*' (Harper no date: 25). While everybody acknowledges that CAT is a very different organisation than it was in the pioneering days and even in the period of Gearchange, CAT's core aim of trying to identify practical solutions for environmental issues has remained a consistent theme. In this sense, although CAT may appear to be a more conventional organisation, it can still be considered a manifestation of practical utopianism.

In the beginning, consensus decision making, the idea of self-sufficiency and egalitarianism were regarded as critical to the quest for solutions to environmental crises. However, in a context in which environmentalism is now mainstream and there is a wide acknowledgement that we are facing a climate emergency, it is a moot point as to whether or not CAT would have continued to be effective if it had retained the consensus decision-making structures. While some understandably believe that CAT has sold its soul to the devil, it perhaps plays an even more significant role now than it ever has. CAT's solution-based discourse is even more critical in the context of the prevailing apocalyptic discourses of the environmental crises that we currently face. While the organisational changes were bought about by a set of very difficult circumstances, it could be argued that these changes were a pragmatic solution to those circumstances and are therefore consistent with CAT's mantra that 'failure is the compost of success'. CAT had worked extremely well for the first three decades of its existence, but circumstance and context had changed considerably by the first decade of the new-millennium and CAT had to find a new way of working. There has to be a balance between pragmatism and idealism in order to achieve its goal to 'inspire, inform and enable humanity to respond to the climate and biodiversity emergency'.

> There has to be an element of common sense and pragmatism in order to get out there and do stuff. You know, and that will always involve an element of compromise. We have to consider if we are ticking all the boxes on the idealism but actually achieving nothing.
>
> (Mick, interview May 2021)

Of course, not everybody agreed that compromise was necessary, or what those compromises might be.

Notes

1 There is almost no information about Tony Williams – what his background was, where he went after leaving the Quarry are unknown.
2 CAT Archives (Box 2/7) Diary 11 July 1974.
3 Although it has to be noted that when CAT was at its busiest in the summer months, the restaurant and the Quarry Shop had to employ casual staff, who were paid an hourly wage, that did not amount to parity.
4 Wright (2010: 244) indicates that in 2007, the Mondragon Cooperative Corporation had roughly 100,000 employees.
5 Later, Liverpool John Moores University validated some of the GSE courses.

References

Borgnäs, K., Eskelinen, T., Perkiö, J. & Warlenius, R. (Eds.) (2015). *The Politics of Ecosocialism: Transforming Welfare*. London: Routledge.

Dufays, F., O'Shea, N., Huybrechts, B. & Nelson, T. (2020). Resisting Colonization: Worker Cooperatives' Conceptualization and Behaviour in a Habermasian Perspective. *Work, Employment and Society* 34(6), 1–20. doi: 10.1177/09500 17019895936

Freeman, J. (2021). The Tyranny of Structurelessness. Available at: www.jofreeman.com/joreen/tyranny.htm (Accessed 5 October 2021).

Harper, P. (no date). Demanding the Impossible: Managing and Mismanaging a Radical Agenda. A Life-Story of the Centre for Alternative Technology. (unpublished paper supplied by the author).

Harper, P. (2016). The Centre for Alternative Technology: An Epitome of Third Sector Values. *NGOs at the Crossroads*. Paper presented at the Oxford Centre for Humanities.

Leach, D.K. (2016). When Freedom is Not an Endless Meeting: A New Look at Efficiency in Consensus-Based Decision Making. *The Sociological Quarterly* 57, 36–70. doi: 0.1111/tsq.12137

Levinson, R., Paraskeva-Hadjichambi, D., Bedsted, B., Manov, B. & Hadjichambis, A.Ch. (2020). Political Dimensions of Environmental Citizenship. In A.Ch. Hadjichambis et al (Eds.), *Conceptualizing Environmental Citizenship for the 21st Century*. Cham, Switzerland: Springer Open, 17–28. Available at: https://link.springer.com/book/10.1007/978-3-030-20249-1

Moscovici, S. & Doise, W. (1994). *Conflict and Consensus: A General Theory of Collective Decisions*. London: Sage.

Putnam, R.D. (2000). *Bowling Alone: The Collapse and Revival of American Community*. New York: Simon Schuster.

Shepherd, A. (2015). *Voices from a Disused Quarry: An Oral History of the Centre for Alternative Technology*. Machynlleth: CAT Publications.

Wright, E.O. (2010). *Envisioning Real Utopias*. London: Verso.

4 Communication, Education and Persuasion

Introduction

This chapter looks at the outreach and educational aspects of the Centre for Alternative Technology (CAT) and analyses its strategies to persuade people to actively address the environmental crises. There is a moral imperative for anyone concerned with the anthropogenic causes of environmental degradation to communicate this concern to others. There is no point in simply withdrawing from the world if it is hurtling towards an ecological disaster. However, simply informing people of the facts is insufficient as it is well established that knowledge of the environmental crises facing humanity rarely leads to people, organisations or institutions taking any action. As CAT's mission statement indicates, any group that is concerned with addressing these issues must also inspire and enable responses to the environmental crises that humanity and the planet are facing. It is essential to persuade the wider public of the viability and achievability of alternatives.

In many ways the focus of CAT has switched from an emphasis on information to an emphasis on inspiration and enabling. This is, at least in part, in response to sustainability moving from the margins to mainstream discourses. It is widely recognised that while most people have at least heard about the environmental catastrophe that threatens the world individuals, industry and governments have not yet been inspired to take sufficient action. The problem facing CAT, and the environmental movement more generally, is that it must be universally inspiring. CAT is in a unique position as over the years it has tried to reach out to a very wide audience – from the tourist who is just looking for something to do on a damp August day to postgraduate students taking a course in some aspect of sustainability at CAT's Graduate School of the Environment (GSE).

Green Rhetoric

Dissemination of ideas was and remains to this day absolutely central to environmentalism generally and CAT in particular. However, dissemination of ideas is a necessary but not sufficient condition for addressing the environmental

DOI: 10.4324/9781003207702-5

catastrophe facing the planet. Environmentalists have to convince people of the accuracy of their analysis and also motivate them to act on it. In other words, there is a rhetorical aspect not just for the CAT project, but also for the environmental movement more generally. CAT has faced a number of challenges that exemplify the difficulty of persuading others of the imperative to address the various aspects of environmental destruction. These challenges are best explored through the rhetorical assertion that a convincing argument is contingent on different aspects of persuasion.

Classical rhetorics suggests that an argument has two aspects: the inartistic (*atechnic*) and the artistic (*entechnic*). The *atechnic* aspects include evidence such as statistics, physical data and so on – the facts so to speak. However, it is well established that factual information on its own is insufficient in convincing people of a particular argument. Aristotle suggested that you also have to consider the art (*techne*) of the person (*rhetor*) making the argument. There are three aspects or different modes of appeal (*pisteis*) to the art of persuasion: the self-presentation of the *rhetor* (*ethos*), the structure and reasoning of the argument itself (*logos*) and the ability to appeal to the emotions of the audience (*pathos*). These are not distinct aspects but work together to create a persuasive argument. For example, no matter how sound the argument is (*logos*), it will not be convincing if the speaker does not appear credible (*ethos*) or the ideas do not emotionally resonate with the audience (*pathos*). Over the years CAT, like many other organisations in the environmental movement, has come to realise that the main issue is not one of facts, but one of behaviour. Consequently, while *logos* remains important in any argument, *ethos* and *pathos* are in many ways more critical in inspiring people to take action on the environmental crisis.

Kenneth Burke observes that if someone is to be persuaded, they must identify with the rhetor: 'A is not identical with his colleague B. But insofar as their interests are joined, A is identified with B'. Burke coins the term 'consubstantiality' to indicate that 'being identified with B, A is "substantially one" with a person other than himself. Yet at the same time he remains unique, an individual locus of motifs' (Burke 1969: 20–1). In the words of M. Jimmie Killingworth and Jacqueline S. Palmer (1992: 8) 'rhetorical appeals propose enlargements of the *we* category'. However, if the adoption of alternative technology (AT) challenges the habitual practices and dispositions of everyday life, then this will militate against extending 'the we category'. The audience may then fail to identify with the AT rhetoric and there will continue to be a sense of 'us' and 'them'. As mainstream and environmental discourses converge there is more likelihood of identification, and hence persuasion.

The challenge for CAT, and the environmental movement more generally, is that the rhetorical strategies must persuade diverse audiences. What makes an argument convincing to a particular audience is not necessarily going to be persuasive for a different group. It is important to frame arguments in different ways for different audiences. Initially CAT did not appear that credible to many in the more mainstream society despite Morgan-Grenville's elite credentials. However, over the years, in part because it has simply survived where many other projects like the Earth Centre have failed, CAT's credibility as a persuasive voice in the environmental movement has increased. CAT's history has contributed to it being taken more seriously by a wider audience, including mainstream political institutions such as the Welsh Government and local authorities. As the Visitor Marketing and Business Development Manager puts it: 'I think that CAT's history gives us credibility in a huge way' (Rob, interview July 2021).

However, CAT's rhetorical style can be a little dry, and not necessarily emotionally appealing, as a focus on technology, scientific evidence and rational argument (*logos*) has tended to take precedence over emotional appeal (*pathos*). It is the site and CAT's historical narrative that have provided the emotional appeal that is lacking in many other aspects of CAT's rhetorical strategy. In both the oral histories and the interviews that I conducted many people refer to the absolute centrality of the Quarry itself without being able to clearly articulate why it is so intrinsic. The narrative of the site has three interconnected strands – The Quarry is a special place, it has been transformed from an industrial wasteland to a verdant habitat that supports a rich biodiversity, and many visitors have been inspired by a visit to the site and have returned as volunteers or to work.

There was a period, mostly in the 1980s and 1990s, when the signage around the site did try to make an emotional appeal. Rather than using very technical diagrams, the two graphic designers responsible for the signage used cartoons.

> How we could get the public inspired to get the grey matter going, inspire them to think about things ... We would have a scenario, for example, with the nuclear family, in the vegetable garden and there would be speech bubbles with the characters discussing the value of growing locally and raising issues about food.
>
> (Graham Preston, Oral History)

However, currently the information signs around the site tend to be very factual and illustrated with photographs and diagrams. This is in part due to the sense that CAT must appear to be increasingly professional, which constrains CAT's rhetorical strategy.

The environmental movement has, until quite recently fallen for the fallacy that people fail to address environmental destruction because they do not have sufficient information. There is a perception that if people are sufficiently informed about the anthropogenic causes of environmental crises, they will act. Consequently, the environmental movement has tended to focus on the *atechnic*. If you use statistical data and/or show images of turtles entangled in plastic netting or starving polar bears adrift on a tiny raft of ice, this will be sufficient to inspire people to act. Of course, the presentation of statistical data as evidence for environmental crises and the selection and composition of these images themselves as visually representing plastic pollution and the climate emergency are themselves rhetorical choices informed by the *techne* of the rhetor. There is a growing consensus that knowledge about the facts of the climate emergency and loss of biodiversity is insufficient to motivate most people to take any action to address the environmental crises that the planet faces (see Norgaard 2011). Renee Lertzman (2015: 3) argues that even 'the assumption that what people care about will naturally translate into actions is erroneous'. If Lertzman is correct, then this is a massive challenge for environmental communication and environmental education. How does one convince people, who seem not only to know the facts of climate change but also to care about the environment, to take action to avert environmental catastrophe?

Much of the information circulated by the environmental movement tends to be focused on the devastating impact of human activity – we are drowning in plastic; there is an alarming loss of biodiversity; and we are already witnessing extreme weather events that have resulted in famine, destruction of homes and loss of life. This is clearly intended to make an emotional appeal that life is so threatened that we must be compelled to act. George Myerson and Yvonne Rydin identify this as 'a dialectic of catastrophe' which they characterise as a rhetoric that suggests 'things are so bad that the solutions *must* be at hand' (Myerson and Rydin 1996: 181).

There is evidence to suggest that being bombarded with information and images of the catastrophic effects of the climate emergency can be counter-productive. There is often an apocalyptic dimension to visual, oral and textual representations of the climate emergency in the public domain, and this can have a disempowering effect. Saffron O'Neill and Sophie Nicholson-Cole observe that 'although shocking, catastrophic, and large-scale representations of the impacts of climate change may well act as an initial hook for people's attention and concern, they clearly do not motivate a sense of personal engagement with the issue' (O'Neill and Nicholson-Cole 2009: 375). They suggest that while images of famine and dead fish in a dried-up lake were ranked as the most salient in representing the importance of climate change, these were 'disempowering at a personal level'. Conversely, images such as fitting a low energy light bulb did not signify the gravity of climate change but were strongly associated with participants' sense that they were 'able to do something about climate change' (O'Neill and Nicholson-Cole 2009: 373). In other words, there is a mismatch between the discourses of a looming eco-apocalypse and the scale of the responses that most people feel is feasible. O'Neill and Nicholson-Cole conclude that spectacular images of environmental devastation 'must be used selectively, with caution and in combination with other kinds of representation to avoid causing denial, apathy, avoidance' (2009: 376).

Positive representations and viable solutions are necessary to balance the eco-apocalyptic discourse. The sociologist John Urry (2011: 122) argues, 'there must be positive alternatives'. Myerson and Rydin call this rhetorical strategy 'the feasible possibility'. They suggest that this rhetorical approach 'expresses the practical question of how to solve the problem … A deliberative stance is adopted: let us review the options, let us see what is possible' (Myerson and Rydin 1996: 187). The rhetoric of feasible possibility encapsulates CAT's stance. Although the understanding of environmental problems has changed over the decades, and therefore the feasible possibilities must be reassessed, this emphasis on practical solutions has been a constant.

There is anecdotal evidence that a visit to CAT can be persuasive. Many people in the oral history archive recall that either they or people that they have encountered have been so inspired by a visit to CAT with school or family that they have returned as volunteers or for a job. There is even some evidence that CAT can have a direct influence on lifestyle choices.

> The most scary letter I ever got was someone who wrote 'we came around the Quarry in August and I went back home to London and as a result I have sold my car'. You think we are responsible in being the trigger in an individual changing in their life to that extent.
>
> (Pete Raine, Oral History)

However, CAT's overall strategy has been to adopt a cumulative approach to communication. The visitor circuit is considered as a sort of precursor to developing a more in-depth knowledge and understanding, and introduces people to feasible possibilities. Only a limited amount of information is appropriate for the signs around site which are supplemented by a small guidebook. The aim is to inspire visitors to delve deeper. Consequently, CAT developed several ways of pointing people to more in-depth information. CAT from the very early days produced information sheets which ranged from tips on organic ways to get rid of slugs to technical information about energy in the home to DIY sheets on how to make a solar water heater from an old radiator. The relevant section in the *Guidebook* from 1995 indicates what information sheets were available for people who required more detail about any aspect of AT. For example, the section 'Energy from the Sun' points to a number of relevant information sheets that were available in the book shop, such as *Solar Energy Fact Sheet* and *Solar Energy Resource Guide*.

CAT also started a publications department and published much more in-depth titles such as *The Whole House Book: Ecological Building Design and Materials* (Borer and Harris 1998). CAT was one of the first publishers to print on recycled paper. Caroline Oakley who was head of publications between 1999 and 2008 stated:

> I think we published some really good books, they're still out there now, people are still buying them and hopefully it has made people make real practical changes to their lives and made a bit of a difference in the world.
>
> (Caroline Oakley, Oral History)

CAT has also always had an information service for people with specific questions about AT. Initially this was quite informal with people phoning or writing in. Rick Dance who was an administrator in the 1980s indicated that every morning there was a pile of post and that the phone was constantly ringing. If inquiries were general, then people were sent the relevant information and asked to cover the cost. If questions were very specific, then the relevant person had to be found. Eventually in the late 1980s CAT developed a specific Information Department and appointed an information officer. CAT also offers a range of weekend courses and it is hoped that a visit to the Quarry will inspire people to take a short course in some aspect of AT.

Informing Others

While recognising that information is important, the emphasis must be on pragmatic solutions that are perceived to be viable and achievable alternatives to environmentally destructive ways of being. When CAT was founded in the mid-1970s, there was very little knowledge and understanding of environmental issues. Except for specific topics, such as saving whales, very local issues or particular disasters like oil spills, the environment was not really a subject of great concern in the public domain.

> The environment was very much a minority thing in the 1980s. It's not like now. I remember we used to keep a scrapbook at CAT. We had *The Guardian* every day and if there was any mention of the environment and you were down in Tea Chest, there would be a great whooping of joy. A little piece about the environment would get cut out and put in this scrapbook. And then another month would go and there would be nothing.
>
> (Rick Dance, Oral History)

Informing others was integral to the CAT project from its inception. In the short document outlining his vision of the nascent project Gerard Morgan-Grenville wrote in 1973 that CAT should '*demonstrate* that the average needs of a domestic situation can be met substantially by alternative technology'. That 'the findings of the experimental research will be *made freely available to all*' and the project would include '*an information office* and *a lecture hall*' (Morgan-Grenville 1973; my emphasis). Consequently in 1975 CAT opened its doors to a paying public. Shortly after opening CAT also started running short courses for the public on various aspects of AT – from wind power to organic gardening. In 1976 Nigel Dudley, who was CAT's first Education Officer, assembled teaching packs for schools on energy, food and various other aspects of AT. Gradually, as the displays were expanded and improved, school groups began to visit the Quarry and guided tours of the site were offered. The focus on education and dissemination has remained core to CAT, culminating in the development of the GSE which offers masters' courses in various aspects of sustainability, and the Zero Carbon Britain Hub and Innovation Lab that advises local governments and other organisations (see Chapter 5).

The Visitor Centre

The public made its way up to the Quarry, almost from the very beginning. Roderick James and Gerard Morgan-Grenville in their meeting, shortly after James was appointed the first director, decided that CAT should be formally opened to the public, and that a small entrance fee would be charged. The primary impetus for this was of course to demonstrate the possibility of leading a low-environmental impact lifestyle. Although Morgan-Grenville had raised the fairly significant capital sum of £20,000 to kick-start the project and was very canny in begging materials from various places, this was clearly insufficient for ongoing running costs. Charging a small entrance fee could also help finance the project. Opening as a visitor centre changed the nature of the project. The main challenge was how best to demonstrate both the problems of the anthropogenic causes of environmental devastation and the potential solutions to a diverse range of visitors, who had different presuppositions and expectancies.

CAT instituted a visitor circuit, and visitors are provided with a guidebook to supplement the explanatory signs around the site, and a plan of the circuit. In recent years CAT has distinguished between a building trail, garden trail and a Zero Carbon Britain Trail. In 2016 CAT developed a more extensive Quarry Trail that takes you past the reservoir, the old quarry face, through woodland, and gives spectacular views across the valley to the mountain Tarren y Gesail. The then CEO Adrian Ramsay indicated: 'The new trail helps bring people closer to nature and local heritage, illustrating the impact that human activity has had on biodiversity and helping visitors understand how we create landscapes that actively benefit nature' (cited in CAT 2016: 4).

The explanatory signs around the site provide information about various aspects of sustainability. Being in Wales, signs also are bi-lingual, which reduces the amount of information that can be included. However, the real challenge is to provide the appropriate level of detail. Issues around the environment are often complex and multi-stranded, and too much or too little detail can fail to inform, inspire and enable.

> There was often a tension between the designers who were trying to keep it on the brief side, knowing the low capacity that people have for absorbing random information and the experts who often tried to include unmanageable levels of detail.
>
> (John, interview February 2021)

Initially the visitor circuit was quite rudimentary.

> In retrospect, it was a pretty embarrassing display. I mean, we almost opened the cupboard and found a hurricane lamp and put a little thermocouple in it, which generated electricity, and that became one of the displays. There

was a sort of fish tank with odd fish in it, and the vegetable garden was, you know, pretty primitive.

> (Roderick James, Oral History)

Liz Todd recollects that there was a sense that they needed 'to put more effort into actually making it an interesting experience for people rather than just people living there and developing AT'. She recalls that at one point her children's Lego bricks were utilised for a display on solar energy.

> The solar roundabout which was made of the children's toys always stands out. It stood in a glass case at the top of the drive and showed how solar energy made things go round. It was very simple. It intrigued people. People had not seen anything like it.
>
> (Liz Todd, Oral History)

People's expectations of a visitor centre were not always matched by the reality of what they encountered. CAT was, at least in its initial pioneering days, an experiment to see what worked and what did not. Consequently, there were often things on-site that did not work to the disappointment of the paying public.

> People would say, well, the windmill is not working. It's not going round. And you think, well, it is a completely windless day. It won't go round. But people were very ready to judge.
>
> (Liz Todd, Oral History)

In addition, the site itself was sometimes, not to put too fine a point on it, rather scruffy and dilapidated, which could undermine its credibility, as Morgan-Grenville would occasionally point out in his missives. The Ogre rota discussed in Chapter 3 was intended to ensure that the visitor centre was sufficiently tidy, that signs were legible and that everything was working.

The Curious and the Pilgrims

This challenge of how much information is optimum is exacerbated by the fact that CAT attracts a diverse range of visitors. The visitors can be roughly divided into two types – the pilgrims and the curious. The pilgrims are already convinced by the imperative to address the environmental crises. In the 1970s, these would be the people who had probably read publications such as Rachel Carson's *Silent Spring* and perhaps subscribed to *Undercurrents*. The pilgrims do not have to be persuaded. Pilgrims often looked for very detailed explanations on things such as the ratio between output and wind speed on an MS2 wind turbine. However, the curious, what one of my informants calls 'the bucket and spade visitor' who just happened to be in the area on holiday, are generally not interested in that level of detail. But lack of an in-depth analysis does not satisfy the pilgrims.

For example, a letter addressed to CAT from a visitor in 2004 observed:

> The introductory video seemed to be pitched at too low a common denominator. I would like to think that most people who come all the way to visit CAT will have a basic knowledge about ecological problems. I wanted to see precisely why problems have escalated over the past quarter decade. For example, why have the Kyoto and Rio Summits failed.[1]

It is clear from this feedback that these visitors who were already well versed in environmental issues and wanted to deepen their knowledge and understanding were deeply disappointed that the depth of information was insufficient to meet their expectations.

Another letter from a visitor also expressed disappointment that the site was not inspiring for their 10-year-old grandson:

> If you cannot fire the imagination of young children, then any effort towards the radical change that we need becomes all the more difficult.[2]

What these letters indicate is that a persuasive argument has to take into account the audience, and with such a heterogeneous audience CAT has to address diverse presuppositions. CAT is trying to reach out to as broad an audience as possible from 'the bucket and spade visitor' to someone from Extinction Rebellion, and from a primary schoolchild to a postgraduate student. Several of my interviewees alluded to the challenge of addressing this multiplicity of audiences and wondered whether this sometimes confused the message.

To accommodate this diverse constituency CAT has always adopted a tiered approach to information. The visitor site itself is geared towards the curious:

> Grazers who will pick up a headline and are much more likely to respond to a graphic or an image than they are over text.
>
> (John, interview February 2021)

CAT also offers guided tours which help put some of the information in context.

> I think when people go on the tours, they see things and they get it. But when people just wander around, they don't get it because it's not obvious.
>
> (Dieter, interview October 2020)

CAT also produced a number of information sheets that provided more in-depth information than can usefully be provided by the explanatory signs which would hopefully address the needs of the pilgrims who might be looking to know more about anything from the technical details of waterpower to hay box cookery. CAT's self-published information sheets also included resource lists, such as manufacturers and suppliers of photovoltaics. Further detailed information can also be found in the various publications stocked in the bookshop. Eventually CAT opened an information desk on-site, but it also was open to providing information about various aspects of AT by phone or post right from the very beginning. More recently this information service has been made available online as well. It offers:

> Free, independent and impartial advice on a wide array of topics relating to sustainability and sustainable living: renewable energy, green building and renovation, water and sewage treatment, organic growing, and more.
>
> (CAT 2022a)

Renewable Energy

Renewable energy, in particular solar, wind and water, is at the heart of the CAT project. Consequently, displays of these have been central to the visitor circuit. However, displaying the AT hardware – wind turbines, water turbines and solar panels – has presented a number of challenges. The main issue for CAT has been that the prominent display of the hardware coupled with the term *alternative technology* can signify that the solution to environmental problems is simply a technical fix, when it must be so much more. As we have seen Peter Harper, who coined the term *alternative technology*, indicated that 'alternative was intended to signify a much wider brief than the mechanical hardware usually implied by the word technology'. He suggests that AT includes all aspects of life as it denotes a sustainable system, and systems are not 'discrete gadgets but god-awful mixtures of hardware, software, organisation and human intelligence' (Harper 2014: 2).

AT can be considered as an ideology (see Dickson 1974), a way of life and as Morgan-Grenville (1973) suggests 'a philosophical alternative to man's obsession with material expansion'. The *Visitors' Guide* from the 1980s suggests that 'to see the philosophy simply in terms of the hardware would miss the underlying reasons for our existence'. Nonetheless, the association with the hardware is difficult to pre-empt. A further problem is that if people only perceive AT in terms of the hardware and technological fixes, this does not seem to be a feasible solution. Installing a small wind or water turbine are not feasible for the majority of people and solar panels were at that time neither particularly efficient nor cost-effective.

On the other hand, there is always going to be a resistance to wider systemic changes. We are creatures of habit shaped in particular cultural contexts. Our habits are shaped by what Elizabeth Shove (2003: 3) has called 'the three Cs – comfort, cleanliness and convenience'. There was a perception that AT and living with renewable technology would entail reduction in comfort and convenience, and possibly even cleanliness. Many people perceived sustainability to be analogous to wearing a hair shirt. Consequently, CAT had to counter this perception, and in the early decades this was a particular challenge.

In keeping with the ethos of self-sufficiency much of the technology had a sort of DIY aspect. For example, the Cretan windmill, which is made with wooden spars and had canvas sales which stood on a prominent ridge that was visible from the road, became a sort of symbol for CAT. As it was primarily used for irrigation in Crete, the design was not particularly effective at generating electricity. The *Visitors' Guide* claims that it was capable of producing '700 watts in a 23 MPH wind', enough for some light, but not enough for domestic appliances like kettles. Furthermore, in high winds someone had to clamber up the steep slate slope and furl the sails to prevent it from being destroyed. CAT published a DIY sheet, which was available in the bookshop for anyone interested in constructing a Cretan-style turbine. The current *Guide* states that the Cretan 'required a great burning of calories as its human operators climbed up and down the hill to unfurl and furl the sails as the wind speed rose and fell' (CAT 2019a:16).

Even in the UK, solar energy can now make a significant contribution to energy demand. There are three main ways of obtaining energy from the sun – passive solar which is about the design and orientation of buildings to make maximum use of the sun's energy; concentrating collectors that focus direct sunlight to heat water; and electric cells, commonly called photovoltaics or PVs, which can directly transform sunlight into electricity. Passive solar is demonstrated in many of CAT's buildings. Solar collectors were also a fairly new technology. CAT had an impressive prototype that was a bank of trough-shaped mirrors that could track the sun's movement and concentrated the sun's energy onto narrow tubes that ran down the centre. One autumn day these just stopped working. Bob Todd realised that because the sun was lower in the sky it had melted the plastic gears. He comments 'whoever had designed it and tested it had obviously never run it under realistic conditions'. Todd continues that 'It is quite an exciting technology and it's now used quite widely in other parts of the world'.

In the early days PVs were not particularly efficient or cost-effective.

> We did quite early on get a small system of PVs. I think it was about 50 watts total or something. It was an interesting exhibit for people to look at, but it didn't really contribute anything significant and most of the modules stopped working pretty quickly when exposed to the Welsh rain for a year or two.
>
> (Bob Todd, Oral History)

Of course, the technology has advanced considerably. CAT now has a large integrated PV roof above the courtyard near the restaurant.[3] When it was first built it was the largest PV array in the UK, covering 112 square metres and capable of generating 13.5 kilowatts – a great deal, more than the 50 watts or so generated by CAT's first PV display. There are now not only solar farms dotted around the countryside but also many individuals have opted to install PVs on the roofs of their homes.

It is impossible to assess how much impact CAT has had on the spread of PVs. Although the development of the technology has made PVs a viable option even in the UK, CAT has had no role in the development of the actual technology. There might have been a handful of people who, over the years, have been directly inspired to invest in solar through a visit to CAT. However, CAT's importance has been both more significant and more diffuse than simply directly motivating a few individuals to invest in solar panels. CAT is a significant actor in an environmental network that, in combination with other environmental actors, has helped amplify a discourse that normalises the use of PVs. While there is now a logical and economic argument (*logos*) for installing PVs, this is not sufficient to persuade. The burgeoning of the discourse on the economic viability of PVs, which includes the increasing visibility of PVs not only on roofs and in fields but also in newspaper articles, advertisements and so on, contributes to a context where the arguments for solar power are more credible (*ethos*) and have a greater resonance for people (*pathos*).

> When people came in the early days around the visitor's circuit, they thought that what we were doing was really radical and really extraordinary.
>
> (Liz Todd, Oral History)

This sense that CAT was radical and extraordinary was a challenge as it was easy for people to dismiss what was advocated around the visitor circuit as not being applicable for the average person. However, the current challenge to persuade people of the urgency to address environmental issues is very different. Unlike in the 1970s–1990s, most people are very familiar with the technology of renewable energy, may well already be convinced about organic gardening and vegetarianism/veganism, and have at least some knowledge about environmental issues. The problem is not so much about overcoming visitors' perception that the ideas about sustainability are too alien to be relevant to their lives, but about challenging a 'so what, I have heard this all before' attitude.

Visitors at the end of the 20th century might have been entranced by a small multi-blade wind turbine with an attached seat that gently goes up and down as the blades rotate in the wind, or a radiator painted matt black that can produce scalding water on a sunny day, but this is no longer the case. In the 1980s, when I was at CAT, no one ever imagined how fast the technology of AT would develop, how large wind turbines would be and how effective PVs could be in the UK climate. It was unimaginable that a major electricity company in the UK would feature wind turbines and solar farms and make the claim that it is helping Britain achieve Net Zero in a television advertisement.[4] The science and technology of AT has developed at such a pace, it is very challenging to maintain an up-to-date display.

Another challenge for the display is that CAT is not simply a tourist attraction, it was also an experiment. In the first decades the hardware of AT was very rudimentary and new prototypes of wind turbines and solar panels were being developed. CAT was interested in both living with AT and assessing how practical some of these prototypes were, and also experimenting with its own designs. There was a tension between these experimental aspects of the early project and CAT as a tourist attraction. Bob Todd recalls getting several prototypes of wind turbines and solar panels which were then incorporated into the display, but many of them failed to work or broke down because of basic flaws in the design. Early wind turbines were very small and only generated a few kilowatts of power. There was a move to develop bigger machines and Todd recollects that 'they were very unreliable. They quickly fell apart or needed a lot of modifications'. This was excellent for experimental purposes, to see what worked and what did not, but not so good for display purposes as many of the day visitors tended to expect displays to be pristine and fully working.

> [The solar wall] was a grand failure, but that was why we did things to see if they would work, because nobody anywhere was doing that … CAT built up a palette of ideas that we later drew upon.
>
> (Pat Borer, Oral History)

CAT claims that it learned a great deal from the experiments with the Cretan and the prototype wind turbines and that these lessons have 'contributed to the development of modern commercial wind turbines' (CAT 2019a: 16).

Sustainable Building

Sustainable buildings that are not only energy efficient to use but are also environmentally friendly to construct have been at the heart of the CAT project since its inception. CAT's website indicates that in our homes 'two-thirds of all energy use is for heating' and 'minimising heat loss will lead to reduced carbon emissions as well as lower fuel bills' (CAT 2021a). Over the years CAT has developed an expertise in environmentally friendly building. For example, CAT was one of the first places to retrofit buildings to make them more sustainable. The slate workers' cottages, which were renovated to provide accommodation for the community, were not on the visitor circuit but could be viewed from a distance by the public. However, there is quite an extensive description of them in the early *Visitors' Guide*, which explains:

> They incorporate energy conserving techniques through the use of insulation, high thermal capacity (which reduces day/night temperature variations) and roof insulation of 4" of glass fibre.
>
> (CAT no date: 8)

However, not all these experiments were successful. Partly because the slate workers' cottages were built into the hillside and they could be cold and damp despite every effort to make them as insulated as possible, according to at least one resident.

At the other end of the site was a more successful project. This is a building that looks like a suburban house and seems rather out of place amongst the vernacular Welsh slate architecture and the predominantly timber framed buildings that characterises most of CAT's additional buildings. This is commonly referred to as the Wates House as it was donated by the building company Wates, although it is labelled as the Conservation House and is now called the Whole Home. It has 450 millimetres (18 inches) of insulation, which made it the most insulated building in the UK. It has quadruple glazing and a heat exchanger to reclaim waste heat. The earlier guidebook describes the building as:

> An attempt to demonstrate how resource consumption in the home can be reduced to a minimum whilst maintaining a very satisfactory level of comfort … The house uses about *one fifth* the energy of a comparable size house … Conservation House is an example of an approach to building which could be applied in urban and suburban areas.
>
> (CAT no date: 27)

The Whole Home is a recognition though much of CAT might seem to be appropriate for rural living, not everything is suited to urban living. After all, the majority of CAT's energy was derived from water, but most people have no access to flowing water with which to drive a turbine. Even erecting a small wind turbine would be problematic for most people, and at the time photovoltaic technology was neither cost-effective nor efficient. The Whole Home is part of CAT's rhetorical strategy to persuade visitors that a sustainable lifestyle does not entail sacrificing the three Cs of comfort, convenience and cleanliness, and returning to some sort of primitive lifestyle that is cold, uncomfortable and arduous, but can be quite consistent with the norms of most people's everyday life.

The ground floor of this building is part of the visitors' circuit and aims to demonstrate the resource and environmental impact of domestic life. For example, there is a typical household noticeboard; however the notes on the board have energy-saving information and tips, such as: 'To make one desktop computer uses 1.8 tonnes of raw materials'; 'Don't replace – upgrade'; and 'Turn it off – don't screen save'.

This attempt to show that CAT is relevant to a more urban setting is reinforced by creating a suburban style garden behind the house. This 'shows the range of fruit and vegetables that could be grown in a domestic organic garden, while still leaving space for a small lawn, an herbaceous border and plenty of flowers' (CAT 1995). In a later guide the garden is described as 'a low maintenance ecological garden that helps the imaginary residents of the Whole Home to live a low carbon lifestyle' (CAT 2019a: 30). The Whole Home also acknowledges that people are often motivated by immediate pragmatic concerns and that environmental issues can seem to be distant and abstract. There is a perception that leading a sustainable lifestyle is expensive and it is therefore important to allay people's concerns. *The Visitors' Guide* from 1980 suggests that building something comparable to the Whole Home would have added about 10% to the cost. However, 'the total space and water heating costs should average about £1.70 per week' (CAT no date: 27).

It is acknowledged in the more recent *Guide* that the technology has advanced significantly, and that the Whole Home would be built very differently today. For example, the windows in the Whole Home are very small to prevent heat loss, and it did not make sufficient use of passive solar energy. The 2019 *Guide* indicates that 'we added the conservatory to make use of passive solar gain. With improvements in window technology and insulation techniques it makes much more sense to let light and heat in with a greater surface area of glass' (CAT 2019a: 25). This approach can be quite persuasive. No one likes being preached at in an imperious way. Being 'talked down to' does not foster a sense of con-substantiality between the speaker and their audience. CAT has successfully adopted a rhetorical style that might be called experiential fallibility, which is persuasive because it 'feels authoritative and definitive, although speaking the language of uncertainty' (Myerson and Rydin 1996: 130). While a few dyed-in-the-wool climate change deniers try and cast doubt by arguing that the science is not a hundred percent certain, the majority are more persuaded by a rhetoric of feasibility than the rhetoric of absolute certitude.

As well as the super-insulated Whole Home, CAT has experimented with a range of sustainable building techniques. For example, from very early on CAT ran self-build courses that were based on the Walter Segal self-build technique. Segal's ideas were very much consistent with CAT's ethos. Segal developed this system of self-build which he claimed can be accomplished by anyone who can saw a straight line and drill a straight hole. Segal's method involves a timber frame, uses stilts so that the building does not require foundation, and uses standard and readily available building materials (see Grahame 2015). In 1987 CAT ran a ten-day self-build course using Segal's methodology, but also ensuring that it incorporated energy efficiency ideas. The basic structure was completed during the course and the CAT builders then finished the construction. However, retrospectively it has been recognised that some of the materials used in the self-build house were not environmentally friendly. For example, the insulation used was Styrofoam which is now recognised as being very damaging to the environment.

With the next major project, the stations for the cliff railway, CAT began to think more about sustainability of materials and issues such as embodied energy (the energy utilised in the actual construction of the building). Cindy Harris, who project-managed the top station recalls:

> What I was interested in were building materials. Utilising alternative materials that were low carbon. For obvious reasons, we were into timber which is the only renewable building material … We started to realise that we were using imported treated timber, and this was not good. So at that point, we stopped using treated timber and started using local timber … It

was the first time that philosophy was translated into a building. And so it really felt like we were progressing that whole development of ecological building materials and of course, it's a beautiful building.

(Cindy Harris, Oral History)

This philosophical approach to building was developed further in the building of the Autonomous Environmental Information Centre (AtEIC), which houses the shop and the drop-in information desk. This building utilised a rammed earth technique. Rammed earth is a very ancient building technique, but not one commonly utilised in the northern hemisphere. This technique involves constructing shuttering, which is then used to contain layers of earth that are compressed in order to build up the walls, and then removing the shuttering. The embodied energy, that is the energy utilised in the construction, is very low as it uses very little or no cement, which has a very high embodied energy and which produces extremely high CO_2 emissions in its manufacture. The 2019 *Guide* indicates that 'the shop was our first foray into architecturally advanced ecological building' (CAT 2019a).

We wanted to show something that was absolutely exquisite. We wanted to do good architecture to show the best of rammed earth.

(Pat Borer, Oral History)

The experience using rammed earth for the shop was developed further in the building of the futuristic Wales Institute for Sustainable Education (WISE) building. This was primarily designed to provide accommodation and teaching spaces for students taking MSc courses at CAT. While it is not open to the general public, there is considerable information about the design and its environmental credentials both on-site and on the website.

The WISE building is designed to have a very low environmental impact in construction and in use. Throughout WISE we used natural materials with a low embodied energy – including timber, earth and hemp. The embodied energy is the total energy used through extraction, transportation and manufacture.

(CAT 2021b)

As much material as possible was sourced locally. For example, the earth for the rammed earth walls was sourced from a quarry less than 50 miles from the site.

The concept was to move away from the timber-framed look, which although beautiful can come across as retrogressive.

The design was intended to get away from people's image of green architecture as very hairy and woolly, curvy and wiggly and grass all over it.

(Pat Borer, Oral History)

David Lea, the architect who worked with Borer in designing the WISE building stated:

> We wanted to find an appropriate architectural language for the time we live in. We were searching very much for that because you do not want to build buildings anymore, which look back all the time to the past.
>
> (David Lea, Oral History)

In other words, WISE was conceived as a sustainable building that uses less timber, looks modern and is both comfortable and functional.

WISE won several architectural awards and was included in both the *Guardian* and *The Telegraph's* top ten best architecture of 2010. An article in *The Architectural Review* describes the WISE building.

> The building proves that energy saving requires no architectural compromise, no necessary meanness, ugliness, or loss of quality. It also shows how energy-saving devices can be integrated without biasing the whole design or spoiling it with add-ons.
>
> (Jones, 2010)

Phil Horton described WISE as an inspiring space.

> Just being able to show architects and engineers what is possible, I think that will be its biggest impact ... We are getting all these people coming through who are learning about architecture and renewable energy and so on. And they're all able to feel inspired by the space that they're in.
>
> (Phil Horton, Oral History)

The buildings therefore are not only functional but are also themselves demonstrations of sustainable architecture. Indeed, it could be said that the various buildings are themselves rhetorical statements. Maria Kanengeiter suggests: 'Architecture can function as a kind of argument in that the viewer can see a building making a claim or an assertion of some kind' (cited in Chryslee, Foss and Ranney 1996: 9). Each of the buildings at CAT not only asserts sustainability in some form, but also constructs the argument differently. The self-build house, for example, articulates the development of the DIY ethos of the early CAT, stating that DIY buildings do not necessarily entail a rustic aesthetic. The Whole Home clearly states that low-environmental lifestyles are not just relevant for a few people living in a remote rural location but are applicable in urban contexts too. The top station with its use of local timber and its stunning views over the Dulas valley emphasises the link between the landscape and architecture.

The WISE building makes the strongest rhetorical statement, and clearly expresses that sustainable building can be beautiful and modern. Myerson and

Rydin (1996:193) suggest that 'overcoming polarity is central in environmental discussion'. The WISE building visually signifies overcoming a number of polarities, such as tradition and modernity. It uses some very traditional construction techniques but is clearly a modern – indeed some people have suggested a futuristic – building. It not only has a modernist architectural style but is also organic. The WISE building appears both natural and constructed. The signs that provide visitors with the details of the building's environmental credentials, both in terms of the energy in construction and use, are clearly an appeal to *logos*. For example, there is a sign that explains that the properties of earth as a building material are 'low embodied energy, natural material, non-toxic, sourced locally, thermal mass'. However, WISE also appeals to and contributes to *ethos*. WISE is a statement that CAT has moved with the times and therefore is a credible voice. Furthermore, as WISE has won several architectural prizes and has been acclaimed in both specialist publications and mainstream media, it has contributed to CAT's esteem and make it less likely that it can simply be dismissed as a bunch of maverick dropouts living on a remote hillside. Finally, as we all know, a strong architectural statement, such as the gothic cathedrals of medieval Europe, can evoke an emotional response (*pathos*). While not everybody will necessarily respond in the same way, entering the cathedral-like space of the circular lecture theatre with its beautiful rammed earth walls can evoke a sense of wonder.

The Garden

While most people associate AT with renewable energy, organic gardening was integral to the display from the very beginning.

> The garden is really important, particularly in terms of climate change and how we farm, how we treat the soil, the way we think about nature and the environment and our role within that … It's not all about how you heat your house or where the electricity comes from. How you relate to the natural world and where food comes from is a very big part of that picture.
> (Petra, interview July 2021)

As the site is largely situated on an old slate quarry, there was very little if any topsoil and mostly the land is on a fairly steep slope, growing anything is a challenge. However, over the years the gardeners have managed to build up the soil and create a verdant oasis. One of the most interesting garden displays demonstrates the process of how the CAT gardeners developed the soil on the otherwise barren slate waste. One part of the garden was divided into four beds: one with soil only, a mix of soil and slate dust in another, slate dust and compost in the third, and finally a mix of soil and compost. The same plants were grown in each of the beds. It is clearly visible that the plants in the two beds with the compost grow far more successfully than those without.

At the time most people thought that it was soil that mattered. But this showed very graphically when planting crops right through these four beds that the compost was the most important.

(Jeremy Light, Oral History)

This display is still there and is one of the best examples of combining experiments with the display. The current *Guide* indicates:

The gardens at CAT are a 'living laboratory'. This means we garden with an experimental attitude, and are always asking questions about how our activities fit into the wider environment … In what ways can gardening contribute to a reduction of carbon emissions, waste, pollution and the depletion of finite resources that have come to characterise modern living?

(CAT 2019a: 28)

Maintenance of the gardens has always depended on volunteers, particularly the long-term volunteers who generally stay for 6 months. There are many people interested in learning about gardening and growing organically and sustainably. CAT has had nearly 50 years of experience which has contributed to its credibility as a place to learn about and get experience of sustainable and organic growing. Many of these volunteers become ambassadors for CAT and are significant nodes in CAT's informal networks of influence.

The gardens have attracted quite a bit of attention from the media, partly because gardening is a bit of a national obsession in the UK. The popular BBC radio show *Gardeners' Question Time* has been recorded at CAT and it has featured on the television show *Gardeners' World* which can have an audience of over 2 million. Perhaps the most significant event was when a small group from CAT were invited to make a small garden at *Gardeners' World Live* in 2002. This event is held every year at the National Exhibition Centre and is represented as 'the ultimate day out for all gardeners' (BBC 2021). The CAT display garden was primarily focused on the UK gardeners' perennial problem of what to do about slugs. The idea was to demonstrate possible ways of dealing with slugs in an environmentally friendly way without recourse to chemical poisons. Allan Shepherd who coordinated the project stated:

Slugs are a minor environmental issue really, but it is a way to connect with people.

(Allan Shepherd, Oral History)

He also suggests that it was a way of going out to people, rather than expecting people to come to the site and thereby reaching out beyond CAT's normal constituency of 'Guardian reading greenies'. The following year CAT organised a garden around the theme of composting, which at the time was not so commonplace. These events enabled CAT to engage visitors to the exhibition with other aspects of sustainability, and perhaps be encouraged to visit and/or pick up a mail order catalogue. Highlights of the exhibition are broadcast on the BBC in prime time. Monty Don, the celebrity gardener who presents *Gardeners' World*, included a short piece about the CAT garden for the show.

> We had had national television coverage before, but this was the first time being on that sort of consumer-focused program.
>
> (Allan Shepherd, Oral History)

Away from the actual visitor circuit there is a small field where most of the vegetables are grown and which supplied the site community and the shared lunches. However, although some visitors thought that CAT was self-sufficient in food, this has never been the case. The field never really produced enough food to supply the restaurant except when there was a glut of a particular crop.[5] Since the demise of the site community the field did supply the restaurant with some salads and vegetables. However, in the last two years the field has been taken over by the Pathway to Farming scheme that was started by one of the garden volunteers. This scheme aspires to train people not only to grow fruit and vegetables, but also to learn about business and marketing. This is clearly consistent with the concept of eating more seasonal and local food, which is a significant part of CAT's sustainability agenda. Pathway to Farming also links CAT more robustly in the local networks. For example, the scheme has directly approached local chefs to see what produce they would be interested in, and specifically grows crops for local food retailers. The idea is to 'build the foundations for a more resilient local food economy that can weather the storms of climate change and global food price fluctuation' (CAT 2021c).

In recent years, CAT has placed more emphasis on the reclamation of the site.

> We talk about the transformational story of how the site used to be a slate waste tip and how through natural regeneration, through careful helping hand of the volunteers it is now this oasis for nature. That is a really lovely story to tell … It links CAT to the wider story of Wales and its story of transformation of the land and its industrial heritage to becoming this important forward thinking country.
>
> (Rob, interview June 2021)

This link to the reclamation of an industrial landscape is particularly apparent on CAT's new Quarry Trail, which takes visitors to the spectacular old quarry face and through naturally regenerating areas and managed woodlands.

The Restaurant and Bookshop

Typically for a visitor attraction there is a restaurant and bookshop. The bookshop was one of the few places in the UK that stocked a comprehensive catalogue of books on renewable energy and other aspects of AT. It was also the place where anyone interested in more in-depth and technical information could buy CAT's own information sheets. The bookshop also sells various environmentally friendly products. As environmental issues became more popular more books about the environment and renewable energy, and eco-friendly products have become more readily available. Selecting the books for the shop was fairly straightforward. For example, it was not appropriate to stock books on gardening which suggested using chemical fertiliser or insecticides. However, the products were often more equivocal. The guiding ethos is that products should be fair-trade and environmentally friendly. Sabrina Cantor, who worked in the shop in the very early days, indicated that whether to stock a particular product was sometimes debated in staff meetings. She stated that:

> The little solar battery chargers are made in China. But that seemed to be the only place where these chargers were made, so sometimes the use of product is put to outweighs other aspects. So, it was quite a juggle.
>
> (Sabrina Cantor, Oral History)

Gradually the shop was able to stock more products like wind-up torches and items made from re-cycled paper or glass, and these started to outsell the books. However, there was always pressure on the bookshop to make a profit which would help fund other aspects of CAT, without compromising the basic principles.

> There were some of the rubbish things for kids we did not need to stock. But then at the same time we were asked to make a big profit in the shop. So if you get something that's very ethically fair trade it can be quite expensive. So you either did not sell it or you did not make as much profit.
>
> (Sabrina Cantor, Oral History)

This points to an ongoing tension at CAT. On the one hand the goal of CAT has always been about identifying solutions to environmental challenges and communicating these effectively to a broad public. On the other hand, as both a business and a charity, CAT has to be financially viable. There was an expectation that the trading aspects of CAT were primarily there to financially support the other aspects of CAT and less about disseminating the sustainability agenda. CAT also built up a highly successful mail-order business. At its height it had a considerable data base and produced a substantial catalogue of books and eco-products. The mail-order business developed into a major source of revenue. However, with competition from Amazon and other internet outlets for books on sustainability and eco-products, the mail-order business began to decline.

The restaurant at CAT was one of the earliest vegetarian places to eat in the UK. Many of the casual visitors in the 1980s and 1990s had not encountered a purely vegetarian menu before. There was a captive audience, so to speak. Certainly, when I was at CAT, I considered the restaurant to be integral to the display and not simply an incidental service. Part of the job was to convince a visitor that vegetarian food is delicious as well as nutritious, when the overriding conception of vegetarianism was that it was a stodgy, bland diet of lentils. Hopefully, we managed to convince some of our customers to at least reduce their meat consumption and to buy a vegetarian recipe book from the bookshop. CAT published a small vegetarian recipe book, *Chop it Cook it Eat It*, based on some of its most popular dishes. While I was at CAT, I produced a small booklet on wholefood baking and an information sheet on the health and environmental benefits of a vegetarian wholefood diet. In many ways the ideas promoted in the gardens and the restaurant were more straightforward for visitors to take home than the more technological aspects of AT.

Tourism

The Dyfi Valley is a popular tourist destination as it is in easy reach of both Snowdonia National Park and the coast. Most of the day visitors, particularly in the summer when the site can get extremely busy, are tourists who just include CAT as a place to visit whilst on vacation. While the day visitors, which at their peak in the late 1990s reached about 90,000 per annum, contributed substantially to CAT's finances, Peter Harper perceived that there was a kind of *laissez-faire* attitude to the site and that visitors would have to take it as it was. Harper comments that he was surprised to discover that although the Quarry was open to the public 'we were not very visitor friendly'. He indicated that he was rather shocked as he considered that CAT was 'in show business'. Here Harper acknowledges the link between how CAT presents itself and how its message resonates with the visitors. Harper also comments that there was sometimes a conflict between what CAT workers personally wanted to do and CAT as a visitor attraction. Many people in the oral history archives indicated that what they appreciated about working at CAT was that no one was their boss, and this compensated for the very low income. Consequently, maintenance was not always high on the list of priorities. There was a sense that people had a right to prioritise their own pet projects, rather than do mundane chores. Peter Harper surmises:

> It is probably fair to say for a lot of the highly skilled people, the knowledge that they were paid very little by the industry norm had an influence on their behaviour … I think that they felt quite unconsciously, look, we've come here, you're jolly lucky to have us. We don't ask for any more money, but we want to be able to play … We don't want to do maintenance, we want to invent things.
>
> (Peter Harper, Oral History)

CAT did not always meet the expectation that some visitors had of a tourist attraction. Overall, the Visitor Centre really divides opinion – some really love it, but others are disappointed. Two typical TripAdvisor reviews read:

> Inspiring day out: It's not all sparkly and shiny like your average tourist destination but that's surely the point. It has heart and soul in a beautiful setting and manages to communicate lots of useful information about environmental issues without being preachy.

> Very disappointed: I expected rustic but not bad disrepair or neglected. I came away after just 50 mins feeling sad and having learned nothing.[6]

CAT differs from most tourist attractions as it not only has to excite, but it also has a mission to inform visitors about environmental issues and inspire them to take some action. As indicated earlier, many people have been inspired by their visit to CAT not only to consider environmental issues, but also to make some changes in their lives.

While some of the hardware of AT perhaps does not have the same fascination as it had in the first decades, CAT does still attract tourists. The Annual Report for 2019 states that, 'day visitors are a key audience for CAT's engagement work' (CAT 2019b). While there is, at the time of writing, a widespread acknowledgement that the display circuit requires major investment, it is also believed that there is enormous potential for the tourist aspect. In particular, CAT can make a major contribution to the idea of sustainable tourism. There is an increasing concern about the environmental damage caused by tourism, and there is a growing movement that is sometimes referred to as sustainable tourism or ecotourism. Tourism is an important sector in the Welsh economy and employs more than 9% of the workforce. In 2015 Wales became the first country to enshrine sustainable development in law with the Well-Being of Future Generations Act. In the wake of this Act, the Welsh government produced a report *Welcome to Wales: Priorities for the Visitor Economy 2020–2025*. This report recognises that a growth in tourism is good for the Welsh economy, however sustainability must be central to the tourist industry.

> In the face of local and global concerns around over-tourism and climate emergency we face new challenges. Our tourism industry is mature and experienced and still has the capacity to grow. But our industry also told us that growth must serve to sustain – not threaten – the things that matter most.
>
> (Llywodraeth Cymru 2020: 4)

With the Welsh government's emphasis on sustainable tourism, CAT has a significant role to play.

We have aspirations in the development project that's coming out to create a sustainable tourism centre of excellence, to help with provision of sustainable tourism related skills. Visit Wales are fully on board with the project and they see CAT as really strategically relevant to their tourism strategy.

(Rob, interview June 2021)

This is definitely an opportunity for CAT as Sheena Carlisle et al (2021: 2) observe that in Wales 'there is a key knowledge deficit on which sustainability skills gaps exist and which sustainability skills are needed in the future in order to implement sustainable tourism'. Rob continues that 'because of its history and reputation CAT has been able to develop a strong relationship with the Welsh Government'. Here Rob identifies that CAT now has a credible reputation and therefore its arguments are more persuasive in the political arena. This is a far cry from the dignitary who accompanied Prince Charles in 1978, recalled by Roderick James, who was overheard muttering 'there is nothing for us here'.

How CAT communicates has also had to change in recent years as concerns about the environment are no longer marginal or regarded as exotic.

In the 1980s people were seeing stuff that could inspire them to do something that they had not thought of. Very few people had actually seen what a photovoltaic panel looks like, so CAT was cutting edge. And that was it was easy to inspire people. What we have now is actually a much more involved argument.

(John, interview, February 2021)

As most people today are more familiar with PVs and more aware of environmental issues it would be reasonable to assume that it would be much easier to inspire people to take action. However, information and awareness have to develop into new ideas and ways of thinking, which in turn need to provoke new ways of acting. Tim Jensen (2019: 7) observes that 'from a rhetorical perspective, the issue is not a matter of *more* or *less* environmental knowledge, but is instead a matter of *how* that knowledge is acted upon' (emphasis in original). The new context can invoke a sense that small personal technical solutions, such as installing PVs or switching from petrol to an electric vehicle, will solve environmental issues. While these are obviously important, the environmental challenges and the solutions are far more complex and involved. In addition the fact that solar and wind farms are now fairly common sights, coupled with the increasing media coverage of international events such as the 26th Conference of Parties (COP26), can lead people to believe that the climate crisis is being dealt with at a government level, and that there is no necessity for individuals to take any action.

Education

In many ways the visitor circuit, to use Roger Kelly's analysis cited earlier, was to 'get people on the first rungs of the ladder'. This initial visit might inspire people to think about reducing their meat consumption, being more organic in their garden or more conscientious about recycling and switching off lights. However, it is also hoped that some of the curious day visitors might be inspired to get on the next rung of the ladder so to speak. The next rung of the ladder offered by CAT are various short courses for the public. Most of these courses ran over a weekend. Hostel-style accommodation and meals in the vegetarian restaurant are included in the price. The early courses were primarily about solar and wind energy.

> There was a huge range of people who attended the courses. From university students through to ministers of small African countries coming to learn about small solar powered projects. These ministers wanted to implement technologies to allow children to read in the dark so they could do their homework at night.
>
> (Jill Whitehead, Oral History)

Joan Randle who took over running the courses in the mid-1990s extended the offer. Not only did CAT run courses on aspects of renewables, organic gardening, and sustainable ways of treating sewage, but it started running young ornithologist holidays for the Royal Society for the Protection of Birds (RSPB), and softer more leisure-orientated courses such as willow weaving as a means to hook people in. Perhaps the most successful courses are the organic gardening courses, as this is something that is inherently interesting for many people in the UK and is something that people can relatively easily incorporate into their lives.[7] In the mid-1980s the self-build courses became quite popular, as there was a general interest in the self-build movement in the UK at that time. More recently courses such as woodland management have become popular. This course is run in a small sustainably managed wood just across the main road from the Quarry. This course, like all other CAT courses, is very practical. The learning outcome of this five-day course is that students will 'gain a comprehensive overview of how to manage woodland sustainably, looking at biodiversity conservation and woodland management skills' (CAT 2021d).

Schools and Colleges

CAT has also been keen to attract school and college groups. In 1982 CAT appointed Joan and Damian Randle – two teachers – to job-share the role of education officer. They quickly established regular school visits, both day visits and short residential stays. Some days there were three or four school groups visiting. Joan Randle estimates that at its height, when it was easier and cheaper for schools to arrange out-of-class activities, there were about 15,000 school children that visited CAT per annum. Schools were given guided tours of the site and often a short class on CAT and environmental issues which would involve practical activities. Joan Randle observes that these brief visits were

often very memorable and inspiring, and she has met many people who were influenced by their brief visits to CAT, some of whom came back to CAT in later life as volunteers or to work (Joan Randle, Oral History).
Current provision for school visits states:

> All our workshops address sustainability across a range of curriculum subjects and many include issues related to global citizenship. Links are made to Science, Geography, Design and Technology, PSE and Citizenship. Embedded in our offer are cross-curriculum links to develop key skills in Literacy and Maths.
>
> (CAT 2021e)

One of the most innovative and impactful of CAT's educational projects are the Eco Cabins. These are two turf-roofed cabins built in 1989 into the hillside with views across the valley. Each cabin can accommodate 18 people. These cabins were designed for residential school groups. The cabins were initially funded through a timeshare scheme. The cabins are completely 'off-grid' and have their own renewable energy supply. There is a prominent display in each of the cabins that shows the power input from the small wind turbine, PVs and a small water turbine that powers the cabins lights and any electrical appliances. Any excess power generated is stored in batteries. Overall power to the cabins is very limited, particularly if there is a dry cloudy spell with little wind. There is never sufficient power for things like hairdryers and if the battery store is depleted there is no energy supply at all. The display also shows the amount of energy being used and the amount of stored power in the batteries. The residential groups can monitor how much power is being used, how much is generated and how much battery reserves they have remaining. Consequently, this experience is intended to raise awareness about switching lights off and to give students the experience of living with and sharing a very limited power supply. Although there are regular flush toilets, each cabin also has a compost toilet. Water has to be bucketed in from a nearby standpipe. Consequently, it involves less work to use the compost toilet. The sewage from the regular toilets is treated using a reed bed system. Any wood used for the wood-burning stoves to heat the cabins and to supplement the solar hot water is weighed and is expected to be replaced by the group. Consequently, groups can monitor their own consumption of energy. The school groups are also expected to do a little gardening during their stay. Damian Randle who was the main instigator of the Eco Cabin project observes:

> Young people staying in the cabins will learn experientially about the question 'What is the impact, on the earth, of a week of my life?' They will learn in the most practical ways possible about renewable energy and storage.
>
> (Randle 1989: 153)

Damian Randle started the journal *Green Teacher* in 1986. He was very keen that this was an international publication, so when he met two teachers with similar concerns on a tour of North America, he invited them to manage subscriptions for the USA and Canada. As it seemed like much of the content was too focused on the UK, with Randle's support a separate North American edition was launched. The idea was that some content would be shared, but that there would also be separate content that would be more relevant for the two regions. Unfortunately, the UK edition folded in 1995, but the North American Edition is still published, albeit in a digital rather than paper form (*Green Teacher* 2021). This again demonstrates the invisible influence CAT has had on the international stage. In 1989 Randle also published the book *Teaching Green*, which not only outlines CAT's philosophical approach to green education, but also provides some practical classroom activities that emphasise the experiential, practical, solution-focused ethos of CAT's educational programme.

CAT's rural location has also been important. Many of the school groups came from the nearby urban conurbation of the West Midlands and many of the schoolchildren had little or no experience of the countryside.

> I remember we had a group from Bradford once and they had never left Bradford. They had never seen clouds on the top of hills and they had never seen rivers and streams with clean water in them.
>
> (Joan Randle, Oral History)

The Graduate School of the Environment

CAT is now involved in teaching a range of postgraduate courses in various aspects of sustainability. The GSE is independent, but the masters' courses are validated by either the University of East London or Liverpool John Moores University. Courses range from a MArch in Sustainable Architecture to MScs on Sustainability, Food and Natural Resources or Sustainability and Behaviour Change. All the courses, with the exception of the MSc in Green Building, have sustainability in the title.

The idea of sustainability has become a commonplace in environmental rhetoric. Derek G. Ross (2017: 8) explains: 'The commonplace as a rhetorical tool works as a sort of argumentative shorthand'. Commonplaces are incorporated into cultural discourses and are condensations of complex and contested phenomenon. Consequently, a simple term such as sustainability can evoke an abundance of meaning and significance. Commonplaces draw upon audiences' prior knowledge and understandings, and therefore will be understood in a wide variety of ways. Nonetheless, although individuals might disagree about the meaning and significance of commonplaces, they will be almost universally resonant. Commonplaces are very useful in titles (indeed the title of this book itself utilises the commonplace sustainability) as they are pithy articulations of content. The idea of sustainability has been around for a long time. However, it is only after the publication of the UN report *Our Common Future* in 1987 that the concept of sustainability begins to circulate extensively in public discourses and develops into a commonplace. Once sustainability has become a commonplace it can be an effective marketing strategy.

While, the establishment of the GSE was controversial, mainly because it broke the principle of pay parity (see Chapter 3) it is arguably one of the most significant aspects of the current CAT project. Unsurprisingly the masters' courses are very practical and solution-oriented. Head of the School, Dr Adrian Watson, writes in the introduction for the prospectus:

> Should you choose to study with us, we aim to equip you with the skills, knowledge and insights to allow you to play your part in creating the sustainable society that is needed now and for the future.
>
> (cited in CAT 2021f: 3)

CAT has developed a reputation that does attract students, and in the last few years there have been more applications than available places. Students can opt for taking the courses by distance learning or they can choose to take a one-week intensive residential study week at CAT as part of each module. This flexible hybrid model of teaching has meant that CAT was ahead of the game in terms of developing its virtual learning environment (VLE) and has not only been able to recruit students who have other commitments but has also been successful in recruiting international students. The GSE students tend to be slightly older than the average MA student in the UK, and most are in employment. The majority of the GSE's students are looking for a career change to working in sustainability, whilst others are already working in the green economy and are studying for a master's degree as part of their career development.

In some ways GSE is both part of yet distinct from the rest of CAT. One of the lecturers suggests:

> The Graduate School felt like it was a very different from the rest of CAT. We got paid on a different pay scale. We had a different mission. We had a different audience.
>
> (Ruth, interview May 2021)

However, the Zero Carbon Britain project provides a unifying theme and has drawn the disparate aspects of CAT together in a more coherent way (see Chapter 5). For example, some of the public short courses now utilise the GSE lecturers. The site itself is also very important for the GSE.

> The site really does demonstrate the idea of practical sustainability. It is a truly inspiring place to come … When a student is on site, they are secluded from the rest of the world.
>
> (Adrian, interview April 2021)

A major aspect of the immersive week at the Quarry is the interactions amongst the students, many of whom have already had some experience of working in some aspect of renewable energy or sustainability.

The WISE building is almost unanimously thought to be an inspiring place to teach about sustainability.

> I think it sets the right mood for lots of events and talks and so on. You are surrounded by all those lovely materials that are all in tune with what you're discussing. If you are having lectures about sustainability or talking about sustainability, it kind of reflects all of those values all around you.
>
> (Phil Horton, Oral History)

The WISE building is very indicative of how far CAT has come from the very rudimentary hostel-style accommodation and the draughty classroom in the strawbale theatre that had previously been used to teach the postgraduate courses.

One of the most popular courses is the MSc Sustainability and Behaviour Change, which lists amongst the learning outcomes that students will 'examine the role of public perceptions of environmental risk management and attitudes to behaviour change' and 'learn about the skills required to facilitate the necessary behavioural changes through successful communication and engagement strategies' (CAT 2022b). The GSE is still one of the only Higher Education (HE) institutions to offer a course on behaviour change and indicates that the GSE 'is ahead of the game'. The popularity of this course reflects the growing realisation that the challenge of addressing environmental issues is not so much to do with the development of new technology, but rather engaging with the issues of why individuals, industry and governments are failing to respond adequately or expeditiously. The development of these courses also reflects that CAT is radical but not too radical.

The success of the GSE is indicative of how far the green economy has developed. In the first decades, CAT was one of the few places where engineers and builders who were interested in renewable energy and concerned about sustainability could find employment. Now, there are plenty of employment opportunities with an increasing imperative for businesses, NGOs and governments to decarbonise their activities. Many of the GSE students go on to form their own eco-businesses and are not only advocates of CAT but also become part of the network of groups and individuals that promote practical solutions to the looming environmental crises that humanity and the planet faces.

> There are little CAT student groups all over the world doing really amazing stuff, and I don't think they would be doing that amazing stuff in the same way if they had not come to CAT.
>
> (Ruth, interview May 2021)

The quarterly CAT members' magazine *Clean Slate* frequently has articles about the successes of the GSE graduates. For example, in the 2021 Winter edition there is a conversation with Owen Morgan an early graduate of the GSE, who now has an award-winning solar panel business in Cambridge. This business was a direct result of his thesis that he wrote as part of his MSc at CAT.

> I used my thesis as an opportunity to discuss and research the microgeneration industry in the UK. I looked at the solar PV market and small-scale renewable energy companies to see what was currently being offered, what was needed and what opportunities were available. This allowed us to create a well-researched business plan.
>
> (Rees 2021: 29)

Digital CAT

Inevitably, CAT now has an online presence. As well as a sophisticated website, CAT has a social media presence on Twitter, Facebook and Instagram. While this digital presence is really important, the site is still really significant.

> The site is in a sense what CAT is. We are more than just a website. The physical place really energises people. And even if people don't visit the site, they know about it. Lots of people have got great websites and are doing fantastic work, but they don't have our history, they don't have our presence, and they don't have our site. That is what makes us different, I think.
>
> (Amanda, interview June 2021)

What is interesting about the website is that are no images of industrial pollution, scorched earth or polar bears balanced perilously on small islands of ice that have become the iconic representations of the environmental crises. The images on the website are almost exclusively of the verdant site itself, wind turbines, students working on a sustainable building project and so on. CAT's Digital Marketing Officer indicated that the main aim of the visual element of the website 'is to showcase the site in the best way possible' (Billy, interview July 2021). These images emphasise ecotopian solutions, rather than the eco-apocalypse.

Like the physical site of the Quarry itself, the website has to resonate with a broad audience.

> There are really very few organizations that I can think of that have such a broad range of audiences and types of people that you are talking to. There is a sort of interesting challenge in terms of how you pitch to a diverse audience, and yet appearing to be a coherent organisation.
>
> (Billy, interview July 2021)

The coherency is achieved through the use of a consistent palette of colours and font. Unsurprisingly the dominant colour on the website is a sage green. Green is the main colour that is associated with the environmental movement as it has connotations of nature and growth. The choice of sage is in the mid-range of saturation. Very high saturated colours appear vulgar and garish and therefore unnatural, whereas low saturation can seem dull, so by using a mid-range saturation as the predominant green emphasises naturalness. This sage green is for example used as the background colour for the text boxes that are positioned in the middle of the screen on many of the pages of CAT's website. These large text boxes have short and pithy statements about various aspects of CAT. The text is white, which stands out clearly on the sage green background.

Sally Hyndman (2016: 44) suggests that 'typefaces set the scene and clue you in to what the words will reveal, independently of what they actually say'. Most of the website utilises a *sans serif* font, which tend to be more legible on screens than serif fonts. *Sans serif* fonts appear less formal, more modern and less cluttered than *serif* fonts, and yet still can convey a sense of authority. The font has a moderate weight and utilises conventions such as a horizontal crossbar on the letter 'e'. Hyndman (2016: 86) calls this style of font 'professional sans serif'. This contrasts with a more rounded font such as *Comic Sans*. Overall, the site has a professional look which engenders not only a sense of trust, but also accessibility.

The text box of the opening page of CAT's website invites the viewer to 'JOIN THE CHANGE'. This exhortation is all in capitals and a considerably larger font size than used in the rest of the text box. The use of uppercase and large fonts is equivalent to adding stress in spoken language, which signifies saliency. When used inappropriately, for example in an email, use of uppercase can be interpreted as the equivalent of shouting. However, the use of capitalisation and large font here, in combination with the actual words, not only invites the viewer to join the change, but also conveys urgency and significance. Below this invitation, still in uppercase, but using a smaller font size, are the words 'Centre for Alternative Technology'. This suggests that by joining with CAT, the viewer will participate in the change. Below this is an explanation in lowercase:

> Radical action is needed if we are to avoid dangerous climate breakdown. CAT offers practical solutions and hands-on learning to help create a zero carbon world.
>
> (CAT 2021g)

The use of a passive voice in the opening sentence and the use of the personal first-person plural pronoun indicates that we are all responsible for both the causes of and the solutions to dangerous climate breakdown. Although CAT is the agent of the second sentence, it is not demanding or commanding but is instead offering help to address the precarious situation that we all are facing. This offer signifies both CAT's expertise and compassion. The tone is one of collaboration rather than of judgement. It is also consistent with the concept of co-creation – that we work together to find solutions. Co-creation is a central idea in CAT's Zero Carbon Innovation Lab, which I will discuss more fully in Chapter 5.

The call for radical action alludes to the utopian aspect of CAT – there are better ways of doing things to ensure the flourishing of humanity and the planet. However, this better way of being does not involve mere tinkering at the edges. A little bit of recycling or replacing petrol and diesel vehicles with electric cars is not sufficient to prevent calamitous climate change. In this sense CAT's solutions can be considered a discourse of 'ecologism' which 'holds that a sustainable and fulfilling existence presupposes radical changes in our relationship with the non-human natural world, and in our mode of social and political life' (Dobson 2007: 3). However, at CAT you can learn about the practical steps that are required to prevent the dangerous climate change. In this sense CAT also provides viable and achievable alternatives to realise the utopian vision. However, whilst the term 'radical' suggests an ideological challenge to the status quo, there is no obvious political dimension to the CAT website. This is at least in part because CAT is an educational charity, but also because proposed solutions to catastrophic climate change must resonate across the political spectrum.

The background to this text box depicts the view from CAT over the verdant woodland to the pronounced peak of Tarren y Gesail on the horizon, which links the exhortation of the text with the natural world. The image suggests that these alternative solutions will preserve the verdant beauty of the Snowdonia landscape. Snowdonia is a designated National Park and is therefore not only an area of outstanding natural beauty and biodiversity, but also one that has been legally recognised. It is a commonplace that these areas should be protected for the benefit of all life – both human and non-human – and that failure to adopt the radical action suggested in the text will result in the decimation of this valued landscape. However, the landscape of the Dyfi valley can be understood as a visual metonym for the planet as a whole. Consequently, the implication is that failure to take radical action threatens not only the cherished landscape of Snowdonia, but also life on earth.

There is also an information service on the website which offers 'free independent and impartial advice' (CAT 2022a). This service provides basic information on a range of topics, such as heat pumps, home composting and reducing water use. The information is clearly written and gives an overview of each of the topics, and also provides links to other websites as well as suggestions of where further details can be obtained. For example, the page on heat pumps explains what a heat pump is; whether it will reduce a household's carbon emissions; how to choose a heat pump, and approximate costs for installation and running. This is very useful basic information for anyone considering installing a heat pump. The section on further information indicates a range of publications and websites that provide more information, including a link to information about a one-day course on heat pumps that CAT will be running later in the year.

As well as courses on-site CAT organises online webinars and online events. These are recorded and many of them are available on the website. For example, at the time of writing upcoming events included a free online Q&A session on Zero Carbon, a one-day online course on compost toilets (which had sold out) and a one-hour forum on organising community climate action with a focus on transport.

Despite some initial reservations by some of the very early pioneers, education and outreach has always been an important aspect of CAT since it first opened its doors to the public in 1975. Like many in the environmental movement, CAT has been challenged to appeal to a diverse audience from a wide variety of backgrounds and with very different presuppositions and knowledge. If the message only resonates with left-leaning, middle-class *Guardian* readers, then the environmental crises that humanity and the planet face will never be adequately addressed.

CAT's location in a relatively remote part of the UK has been a sort of mixed blessing. Obviously, in UK terms, CAT is relatively difficult to actually travel to. However, CAT is located in a popular tourist destination and has been able to attract visitors to the site who are not necessarily particularly concerned about environmental issues. It is difficult to assess what impact CAT has had on the casual day visitors. While many of these visitors simply perceive CAT as a nice day out with the kids, and may view CAT as an interesting, but rather curious anomaly, there are many others who find the place inspiring. There is plenty of evidence from visitor feedback and comments on TripAdvisor that CAT has inspired people to get on the first rung of the ladder. This might be something as simple as purchasing a vegetarian cookbook or a more radical action such as getting rid of one's car. There is also plenty of anecdotal evidence that people have been inspired by a visit to find out more by enrolling on a course or returning as a volunteer.

Another mixed blessing for CAT's outreach and educational programme is that environmental issues and sustainability are now thoroughly embedded in mainstream discourse. This has entailed a twofold shift. First, there has been the imperative to move from an information deficit model of communication to one that is more focused on overcoming barriers to change. Second, there has been the need to move from the margins of society to a more central role. These responses to the changing context are very apparent in CAT's Zero Carbon Britain project, which is the focus of the following chapter.

Notes

1 Letters held in the CAT Archive at the National Library of Wales Box 9/10.
2 Letters held in the CAT Archive at the National Library of Wales Box 9/10.
3 Integrated roofs are where the PVs are also the roof tiles, rather than simply panels installed on top of an existing roof.
4 See www.youtube.com/watch?v=xdyEOMCNTf8

5 When I was running the restaurant Roger, the main gardener, or one of his volunteers, would daily pick a mix bowl of some of his salads including rocket, mizuna, lambs lettuce (which were not really eaten back in the 1980s) and nasturtium flowers – which we always had on the counter for all the customers to help themselves to get a taste of really fresh and what were then unusual salad leaves that can easily be grown in the UK. I remember that one radio journalist who was doing a short feature about CAT commented that we 'served weeds in the restaurant'.
6 These TripAdvisor reviews were published in October 2021. There was no lockdown at this point because of COVID-19. Nonetheless, like many tourist attractions, CAT had suffered because of COVID-19. However, these reviews are pretty representative of the diverging views that visitors have of CAT, irrespective of the impact of COVID-19.
7 We did run a few vegetarian cookery courses whilst I was at CAT, but this was challenging as we had to also cater for the general public at the same time, and space in the restaurant kitchen was very limited.

References

BBC. (2021). *BBC Gardeners' World Live*. Available at: www.bbcgardenersworldlive.com (Accessed 19 November 2021).

Borer, P., & Harris, C. (1998). *The Whole House Book: Ecological Building Design and Building Materials*. Machynlleth: CAT Publications.

Burke, K. (1969). *A Rhetoric of Motives*. Berkley: University of California Press.

Carlisle, S., Zaki, K., Ahamed, M., Dixey, L., & McLoughlin, E. (2021). The Imperative to Address Sustainability Skills Gaps in Tourism in Wales. *Sustainability* 13(3), 1161. doi: 10.3390/su13031161

Centre for Alternative Technology (CAT). (no date). *Visitors' Guide*. Machynlleth: CAT Publications.

Centre for Alternative Technology (CAT). (1995). *Official Guide Book*. Machynlleth: CAT Publications.

Centre for Alternative Technology (CAT). (2016). CAT News: New CAT Trail Explores Wildlife and Local History. *Clean Slate* 102, 4–5.

Centre for Alternative Technology (CAT). (2019a). *Guide*. Machynlleth: CAT Publications.

Centre for Alternative Technology (CAT). (2019b). *Annual Report and Financial Statement*. Available at: https://cat.org.uk/strategy-and-governance (Accessed 19 November 2021).

Centre for Alternative Technology (CAT). (2021a). *Low Energy Buildings*. Available at: https://cat.org.uk/info-%20%20resources/free-information-service/building/low-energy-buildings (Accessed 29 October 2021).

Centre for Alternative Technology (CAT). (2021b). *The WISE Building*. Available at: https://cat.org.uk/info-resources/free-information-service/building/the-wise-building (Accessed 15 November 2021).

Centre for Alternative Technology (CAT). (2021c). *Pathways to Farming: Strengthening the Local Food Economy*. Available at: https://cat.org.uk/pathways-to-farming-strengthening-the- local-food-economy (Accessed 3 November 2021).

Centre for Alternative Technology (CAT). (2021d). *Sustainable Woodland Management*. Available at: https://cat.org.uk/events/sustainable-woodland-management (Accessed 23 November 2021).

Centre for Alternative Technology (CAT). (2021e). *School Visits*. Available at: https://cat. org.uk/school-visits (Accessed 5 November 2021).

Centre for Alternative Technology (CAT). (2021f). *Graduate School of the Environment: Prospectus*. Available at: https://cat.org.uk/courses-and-training/gradu ate-school/prospectus (Accessed 16 January 2022).

Centre for Alternative Technology (CAT). (2021g). *Centre for Alternative Technology: Home*. Available at: https://cat.org.uk (Accessed 18 October 2021).

Centre for Alternative Technology (CAT). (2022a). *Free Information Service*. Available at: https://cat.org.uk/info-resources/free-information-service (Accessed 14 January 2022).

Centre for Alternative Technology (CAT). (2022b). *MSc Sustainability and Behaviour Change*. Available at: https://cat.org.uk/courses-and-training/graduate-school/cour ses/sustainability-and-behaviour-change (Accessed 17 January 2022).

Chryslee, G.J., Foss, S.K., & Ranney A.L. (1996). The Construction of Claims in Visual Argumentation: An Exploration. *Visual Communication Quarterly* 3, 9–13. doi: 10.1080/15551399609363319

Dickson, D. (1974). *Alternative Technology and the Politics of Technical Change*. London: Fontana.

Dobson, A. (2007). *Green Political Thought*. London: Routledge.

Grahame, A. (2015). This Isn't at all Like London: Life in Walter Segal's Self-Build 'Anarchist' Estate. *The Guardian* (16 September). Available at: www.theguardian. com/cities/2015/sep/16/anarchism-community-walter-segal-   self-build-south-london-estate (Accessed 22 October 2021).

Green Teacher. (2021). *A Brief History of Green Teacher*. Available at: http://greenteacher. com/magazine/about-us (Accessed 5 November 2021)

Harper, P. (2014). *Songs of Experience: Four Decades of Successes and Failures at the Centre for Alternative Technology*. Originally published in *Whole Earth* in 2002 – updated in 2014 (Unpublished paper supplied by the author).

Hyndman, S. (2016). *Why Fonts Matter*. London: Virgin Books.

Jensen, T. (2019). *Ecologies of Guilt in Environmental Rhetorics*. Cham: Palgrave Pivot.

Jones, P.B. (2010). Wales Institute for Sustainable Education by David Lea and Pat Borer. *The Architectural Review* (21 December). Available at: www.architectural-review. com/today/wales-institute-for-sustainable-education-by-david-lea-and-pat-borer-machynlleth-wales-uk (Accessed 17 October 2021).

Killingworth, M.J., & Palmer, J.S. (1992). *Ecospeak: Rhetoric and Politics in America*. Carbondale: Southern Illinois University Press.

Lertzman, R. (2015). *Environmental Melancholia: Psychoanalytic Dimensions of Engagement*. London: Routledge.

Llywodraeth Cymru. (2020). *Welcome to Wales: Priorities for the Visitor Economy 2020– 2025*. Available at: https://businesswales.gov.wales/tourism/welcome-wales-priorit ies-visitor-economy-2020-2025 (Accessed 11 November 2021).

Morgan-Grenville, G. (1973). *Society for Environmental Improvement: The National Centre for Alternative Technology*. Unpublished paper available at The Centre for Alternative Technology Archive held at the National Library of Wales – reference Box 1/12.

Myerson, G., & Rydin, Y. (1996). *The Language of the Environment: A New Rhetoric*. London: UCL Press.

Norgaard, K. M. (2011). *Climate Change, Emotions and Everyday Life*. Cambridge, MA: MIT Press.

O'Neill, S., & Nicholson-Cole, S. (2009). '"Fear Won't Do It": Promoting Positive Engagement with Climate Change Through Visual and Iconic Representations. *Science Communication* 30(3), 355–79. doi: 10.1177/1075547008329201

Randle, D. (1989). *Teaching Green: A Parent's Guide to Education for Life on Earth.* London: Green Print.

Rees, A. (2021). CAT Conversations: Owen Morgan – Award Winning Renewable Energy Entrepreneur. *Clean Slate* 122 (Winter 2021): 28–9.

Ross, D.G. (2017). Introduction, in D. G. Ross (Ed.) *Topic-Driven Environmental Rhetoric.* London: Routledge, 1–21.

Shove, E. (2003). *Comfort, Cleanliness and Convenience: The Social Organization of Normality.* Oxford: Berg.

Urry, J. (2011). *Climate Change and Society.* Cambridge: Polity Press.

5 Zero Carbon Britain

Introduction

The focus of this chapter is on CAT's Zero Carbon Britain (ZCB) project. In 2007, CAT published the report *Zero Carbon Britain: An Alternative Energy Strategy*. This has been followed by six further reports; the founding of a ZCB Innovation Lab to bring together multiple stakeholders to address the challenges of the climate emergency; the development of a Zero Carbon trail at the Quarry; and an online Zero Carbon Resource Hub with links to a variety of other groups and organisations addressing climate change. CAT also runs regular seminars and webinars on achieving zero carbon; this has become particularly significant since many local councils have declared a climate emergency and there is increasing pressure on businesses and industry to reduce their carbon footprint.

This chapter investigates how CAT has responded to two interconnected factors. First, the recognition that climate change caused by greenhouse gas emissions is not only the most significant environmental issue of the day but is also far more serious and pressing than anyone had hitherto thought. Second, that by the early 21st century, environmental issues were embedded in political, economic, social and cultural discourses. At the beginning of the 21st century, the focus of concerns about the environment were both very different from and yet had a continuity with the issues that were of concern to CAT when it was founded in the mid-1970s. The climate emergency was not recognised as an issue until the 1990s, but throughout its history, CAT has been concerned with identifying solutions to the anthropogenic causes of environmental degradation.

From the Margins to the Mainstream

In some senses, CAT could be considered a success story – environmental issues have moved from the margins to the mainstream; renewable energy is now both technically feasible and politically acceptable; the Montreal Protocol 1987, to phase out chemicals that deplete the ozone layer, has been ratified by all 198 nations of the United Nations (UN), and recycling has become commonplace. This is not to argue that CAT was directly responsible for making these

DOI: 10.4324/9781003207702-6

changes, but I would agree with the claim that 'CAT is much more influential than it appears' (Rob, interview June 2021). CAT's influence is at least in part due to its history and reputation. This history and reputation have meant that CAT is now perceived as a credible voice, not only on the margins but also by local and national governments, and mainstream organisations that now recognise that sustainability must play a central role in their activities. There is an urgency to decarbonise every aspect of life if we have any chance to prevent a catastrophic rise in global temperatures. CAT is well-placed to play a critical role in this, as it has been investigating the possible solutions to the climate emergency through its ZCB project. This major and ongoing project has itself been informed by CAT's previous decades of experiences and experimentation.

There has been an increasing awareness that human activity has affected the environment, and in particular the climate, in a far more alarming way than had previously been realised. An awareness of climate change is not new. Climate change science and the recognition that carbon dioxide (CO_2) and other greenhouse gases have the potential to increase the temperature of the earth have been known about since the middle of the 19th century. By the end of the 19th century, Svante Arrhenius had concluded that the CO_2 emissions from burning fossil fuels could cause global warming. However, it was not until the 1980s that it was realised more widely that annual global temperature was potentially catastrophically rising, and that this was primarily due to human activity. In 1988, the UN Intergovernmental Panel on Climate Change (IPCC) was founded: Its aim was:

> To prepare a comprehensive review and recommendations with respect to the state of knowledge of the science of climate change; the social and economic impact of climate change, and potential response strategies.
>
> (IPCC 2021)

In 1990, the IPCC produced its first assessment report on climate change. In the same year, The UK Conservative Government published *This Common Inheritance: Britain's Environmental Strategy*. G. Kearns (1991: 363) suggests that this White Paper was the first real engagement with green ideas by the Conservative Party, which they presented as 'a major new departure in politics'. These two reports signify that environmental issues, and in particular climate change, were now regarded as a serious consideration in the political domain at national and global levels. However, although by the mid-1990s climate change was well established on the political agenda, it was not prevalent in the media. While specific environmental disasters, such as the *Torrey Canyon* oil spill in 1967, were deemed newsworthy, climate change did not seem to conform to the news values of relevance and topicality (see Brighton and Foy 2007). Editors perceived climate change to be too vague, abstract and distant to be of interest to their viewers and readers.

Monica Djerf-Pierre, in her quantitative study of the coverage of environmental issues in Swedish public service television, notes that out of the 23 categories of environmental stories that she identifies, there is no news coverage of climate change/global warming between 1961 and 1985. Then between 1986 and 1990 one percent of environmental news stories are about global warming. However, in the period between 2006 and 2010, it jumps to 21 per-cent of environmental stories (Djerf-Pierre 2012: 297). In other words, it is not until the early 2000s that climate change has been considered newsworthy. In part, this is because the effects of climate change have become more visible with images such as glacial ice collapsing into the Arctic, starving polar bears drifting on ice floes and parched earth becoming iconic representations of the problem. In more recent years, extreme events such as devastating floods and cataclysmic fires have been directly linked to climate change and therefore satisfy the news values of topicality and relevance.

Many media scholars suggest what has been called the agenda-setting theory. This theory was summarised by Bernard Cohen, who observed that 'the press may not be successful much of the time in telling people what to think, but it is stunningly successful in telling its readers what to think about' (cited in Lowery and DeFleur 1995: 267). Consequently, climate change is now not only a sig-nificant issue in the political and media domains but is also well incorporated into wider public discourses. Anders Hansen notes that many studies suggest that the agenda-setting role of media has been particularly strong on environ-mental issues, as environmental issues are 'unobtrusive', and are 'often not easily observed or experienced first-hand' (Hansen 2019: 159). Although Hansen acknowledges that climate change is not always unobtrusive as we can experi-ence its effects first-hand, it is only relatively recently that natural disasters have been directly linked with climate change.

> It is not so long ago that a whole host of natural phenomena such as flooding, hurricanes, droughts, hot summers, and so on, would *not* auto-matically have triggered references to global warming or climate change in the way that has now become more or less customary.
>
> (Hansen 2019: 159)

In 2000, The Royal Commission on Environmental Pollution published a report on climate change, the significance of which is noted by Allen:

> They are an established, academically credible body. They were not a group of radicals. They were not Greenpeace. And so a lot of mainstream people then began thinking about this.
>
> (Paul Allen, Oral History)

Perhaps it was Al Gore's film, *An Inconvenient Truth*, released in 2006 that both signalled and contributed to a growing public awareness of the threat of global warming. It was screened at several international film festivals including Cannes, was one of the highest grossing documentary films and won two

Oscars (Best Documentary Feature and Best Original Song) at the Academy Awards for 2006.

There were similarities between the context at the turn of the millennium and the mid-1970s, when CAT was founded. Both can be characterised by an apocalyptic sensibility. In the 1970s, the overriding concerns were the finitude of the earth's resources, signified by *The Limits to Growth*, and the destructive nature of industrialisation, exemplified by *Silent Spring*. It is not that these environmental problems have been solved, but that they have become overshadowed by the looming climate apocalypse. There are two major, albeit interrelated, differences between the 1970s and the early 2000s – one that pertains specifically to CAT and the other that relates to the wider context. First, by the early 2000s, CAT was beginning to gain a positive reputation and could no longer simply be dismissed as a group of mavericks living in a remote rural area who were out of touch with the reality of most people's lives. Second, although in the 1970s most people were not particularly aware or concerned about environmental issues and only relatively few shared a sense of an eco-apocalypse, by the end of the first decade of the 21st century it was difficult to avoid a sense of environmental doom, with a constant barrage of images in the media of melting ice, parched earth, floods and wildfires. As Frederick Buell observes, 'a sense of unresolved, perhaps unresolvable, environmental crisis has become part of people's normality today'. While he suggests that 'faith in effective action has diminished' (Buell 2004: xvii), practical utopian visions, as exemplified by CAT, are a necessary countervailing discourse that are as much a psychological necessity as a quest for concrete pragmatic solutions.

By the mid-1990s, CAT recognised that climate change was a significant environmental issue. In the 1995 *Official Guide*, global warming is listed first in a list of eight *Global Enemies*. In typical CAT style, as well as identifying 'the enemy', the *Official Guide* offers some practical solutions:

> Using less fossil fuel is the easiest way for us individually to make a difference. This means utilising energy conservation and more renewables, and it means more public transport or bikes and fewer journeys using private motor vehicles and freight lorries.
>
> (CAT 1995: 30)

This identifies both systemic changes, such as increasing use of renewable energy, and individual changes, like reducing our fossil fuel use by energy conservation and making less use of private cars. However, despite identifying global warming as a significant threat to the environment, it was not particularly high up on CAT's agenda. This all changes in 2007 when CAT produces the first report *Zero Carbon Britain: An Alternative Energy Strategy*. While it is recognised that there are other environmental challenges, these are considered to be either directly related to climate change or to pale into insignificance compared to the threat of the climate emergency.

> We could spend a lot of time developing and researching things that aren't
> urgent – problems that are lesser problems, that could wait for 25 or 30
> years before we have to solve them, so climate change moved to the centre.
>
> (Paul Allen, Oral History)

The realisation that climate change was a wicked problem that required an
urgent response entailed that the primacy of ends had to take precedence
over the purity of means. Consequently, in the early 2000s, several people at
CAT began to accept the urgency of addressing climate change. Peter Harper
suggested that in the 1990s there was a fairly *laissez-faire* attitude about climate
change.

> We saw the climate change problems coming in, but you could well believe
> you had 50 or 100 years to sort it out.
>
> (Peter Harper, Oral History)

In the mid-1990s, Harper calculated that we needed to achieve an 80 percent
reduction in carbon emissions to avoid catastrophic climate change. Harper
indicated that this was more or less what CAT was achieving at the time, by
generating the majority of its power using renewables.

> I was supremely relaxed. I thought, okay, we are going to crack this. It is
> okay. We have got a model that we can talk about. And I did. And people
> said, all right, well, that makes a lot of sense. We could do that because we
> were actually doing it at CAT.
>
> (Peter Harper, Oral History)

Another shift in focus can also be identified in CAT's discourses concomitant
with the increasing awareness of the seriousness and urgency of addressing the
climate emergency. This was a move away from more individual and personal
lifestyle choices to an emphasis on systemic changes. This is not an either-or
scenario, as almost from the beginning CAT recognised some necessity for
systemic changes. Nor does this shift in focus suggest that the challenge to
address the climate emergency precludes individual changes. The ZCB, as the
name suggests, is focused on the national level with an emphasis on political,
economic and systemic changes. However, ZCB does not ignore the fact that
the climate emergency is a global issue and calls for 'effective international as
well as national policies' (Helweg-Larsen and Bull 2007: 16). The third ZCB
report, *Raising Ambition: Zero Carbon Scenarios from Across the World*, published in
2018 focuses more on global responses to climate change. In this report, CAT
collated a number of case studies of decarbonisation projects from around the
world (Allen and Bottoms 2018). This reinforces the claim that net zero is a
viable and achievable aspiration at a global level.

Before looking at ZCB in more detail, I will briefly discuss its precursor *The
Alternative Energy Strategy*, which is widely acknowledged to be the inspiration

for ZCB. The Executive summary of the first report states that ZCB 'has its origins in *An Alternative Energy Strategy for the UK*'. This 'document proposed the then heretical view that energy growth should slow down and then contract' (Helweg-Larsen and Bull 2007: 5).

An Alternative Energy Strategy for the United Kingdom

The underlying ethos of the very early days of CAT was informed by the idea of self-sufficiency. However, self-sufficiency is impractical for most people, and it was increasingly realised that it was not going to address the scale of the environmental problems faced by humanity and the planet.

> One of the criticisms we had from visitors and other people was that it was all terribly small scale and wasn't really going to make any impact on a country like Britain.
>
> (Bob Todd, Oral History)

Bob Todd got together with a small group of scientists and an economist who had an interest in renewable energy to assess whether the solutions proposed by CAT could be scaled up to a national level. They produced a small booklet *An Alternative Energy Strategy for the United Kingdom*, which CAT published in 1977. The context for the study was *The Limits to Growth* that had been published five years before, the oil crisis of 1973 and the concept of peak oil. Consequently, the depletion of resources is the primary focus of *An Alternative Energy Strategy*.

> We managed to piece together this scenario that suggested that unlike the British government's predictions, we with a bit of effort could keep Britain's energy consumption probably relatively constant and possibly even reduce it.
>
> (Bob Todd, Oral History)

Todd and his co-authors conclude that the UK could meet its energy requirements using a combination of renewables and some coal.[1] This could be achieved without having to resort to nuclear power, without any major impact on lifestyle and 'without serious detriment to the environment' (Todd and Alty 1977: 27).

Todd suggests that the proposals were quite radical for the time. In the report, the authors observe that the trend in determining energy policy was to extrapolate from demand, which has historically increased since the Industrial Revolution, then calculating how that increasing demand could be met in the future. Todd and his co-authors suggest, given the finitude of fossil fuels and the unacceptable dangers of nuclear power, that energy policy should be determined by the availability of energy supply rather than by exponentially increasing the demand that ultimately could not be met. Consequently, rather than suggesting ever increasing energy consumption, as the quote above indicates, energy consumption in the UK should be kept unchanged or even decreased. Despite its challenge to the accepted 'wisdom' of the time, *An Alternative Energy Strategy for the United Kingdom* garnered some political interest. Todd suggests that it had some credibility because there was a quantitative aspect to their analysis.

> In the environmental movement people were saying 'this is what we should be doing'. And it was kind of all talk. What we actually did was to try and put some numbers on it, and I think because of that, it did have quite an impact.
>
> (Bob Todd, Oral History)

The style of *An Alternative Energy Strategy* is informed by the academic backgrounds in science and economics of the authors and is underpinned by logical positivism, the idea that there can be an objective reality that exists independently of the observer and can be quantified. While clearly there is an argument throughout the text that is informed by the subjective values of the authors, this argument appears more credible because it is derived from numerical data. The quantitative aspect of the text also appears more plausible as five of the co-authors have PhDs.

Logical positivism makes a distinction between facts and values. Facts are often equated with quantitative data and it is often assumed that you cannot argue with numbers. This has led to two interrelated ideas. First, scientists provide the statistical information (facts) on which policy makers can base their decisions. Second, if an argument is based on numerical data, it is more likely to be persuasive as it is derived from 'objective facts' that are empirically verifiable. As Sıdıka Başçi and Nadia Hassan (2020: 76) observe, 'one of the most effective and powerful among modern forms of rhetoric is the use of statistics'. This is not to suggest that statistical data are necessarily wrong, but that 'nearly all of the numbers currently in use by statisticians and econometricians are mixtures of facts and values' (Başçi and Hassan 2020: 79). There is clearly a polemical aspect to *An Alternative Energy Strategy*. The core argument suggests that renewable energy sources are inexhaustible and clean, but fossil fuels are finite and polluting. Therefore, we should minimise the use of fossil fuels and maximise our use of renewables to power our homes and businesses. However, the statistical data on which the argument is based are itself not value free.

An Alternative Energy Strategy calculates the potential capacity of a variety of renewable energy sources utilising statistics and numerical calculations. For example, the report indicates 'these calculations suggest that taken together, solar installations for water, space heating could provide 198 million MWh of useful energy in 2025' (Todd and Alty 1977: 6). The utilisation of quantitative data and minimal poetic flourishes suggest an objective scientific and evidence-based analysis, rather than unfounded speculation. This sense of 'scientific objectivity' is reinforced by a predominant use of the passive voice and nominalisation. The use of the passive voice obscures the presence of an active agent. Nominalisation, when a verb is used as a noun, further disguises agency.

These linguistic conventions are frequently used in scientific writing to emphasise a sense of empiricism, objectivity and rationality. For example, the report suggests 'the growth of energy consumption has been accompanied by an increasing realisation that the resources available to the UK are limited' (Todd and Alty 1977: 1). By obscuring the agent of growing consumption, the authors indicate the factual nature of the claims. The lack of agency also attenuates any blame − it is no one's fault that consumption is on the increase. Likewise, lack of an agent in the realisation of the limits of resources is also indicative that it is a factual and universal realisation. If, for example, the report suggested that 'we realise that resources are limited', this can potentially signify something different as the first-person plural pronoun 'we' is ambiguous. 'We' can be read as both inclusive and exclusive. 'We' can signify a universal inclusiveness, but it can indicate simply 'we' the authors and appear to exclude the reader and therefore not be as persuasive as the use of nominalisation and the use of the passive voice.

While *An Alternative Energy Strategy* is presented as based on objective, quantitative data, it also advocates a definite political agenda that was at odds with the ethos of the time. For example, it states that 'the conventional belief that happiness is proportional to energy consumption would need to be displaced' (Todd and Alty 1977: 4). In other words, it challenges assumed wisdom that gross domestic product (GDP) is a measure of success, and that accumulation of material goods is the path to happiness.

In 1977, environmental issues were seen as marginal at best and there was still a perception that the environmental movement was inherently countercultural. This perception militates against the argument, no matter how logical and grounded in solid numerical evidence. Bob Todd recalls that although he was invited to present to various government committees and think tanks, he sometimes received quite a hostile reception.

> I was ushered into a big meeting, where lots of important looking people were sitting around a big posh table. The chairman opened the meeting by welcoming me and saying, 'oh, I was expecting a hairshirt and sandals'. It was quite disparaging in a way.
>
> (Bob Todd, Oral History)

Clearly, this chairman not only identified concern for the environment with the counterculture, but his reference to a 'hairshirt' identifies AT with austerity and renouncing the comfort and conveniences of modern living. This is a major barrier to change, which to some extent still applies. However, the authors of *An Alternative Energy Strategy* acknowledge the 'hairshirt and sandals' perception that many people have of the environmental movement, and argue that reducing energy consumption and increasing the use of renewable energy do not entail sacrificing the comfort and convenience of modern life.

If it is accepted that humanity and the planet faced a serious environmental plight, the other challenge is to overcome the hubris of what John Dryzek calls a Promethean Response, which is an 'unlimited confidence in the ability of humans and their technologies to overcome any problems' (Dryzek 2005: 52). The Promethean asserts that human ingenuity will ultimately solve the problem of the limits to growth, and therefore, there is no need to curtail economic activity. For example, as Bob Todd notes in his oral history interview, nuclear energy – which can be construed as a Promethean Response – was considered by many to be the panacea for any future oil crisis. However, a substantial section in *An Alternative Energy Strategy* argues there are too many inherent risks involved in nuclear energy. The authors corroborate their stance against nuclear energy by referring to The Royal Commission on Environmental Pollution Sixth Report of 1976 that lists a range of hazards associated with nuclear power. There are references to other research papers and reports throughout the text. Referring to other sources and research is not only standard academic practice, but also contributes to the credibility of the argument. This credibility is emphasised by a statement of endorsement in the Preface by Sir Martin Ryle, who was at the time the Astronomer Royal, someone at the heart of the establishment and a highly reputed scientist.

The style of *An Alternative Energy Strategy* raises the questions, as do the later ZCB reports, of what sort of text it is and who is the intended readership. While all the authors have an academic background and use some of the conventions of academic writing, their report is not an academic paper and it is not intended for a specialist academic readership. The text minimises the use of specialist technical terminology, it was published by CAT and not in a recognised academic journal and therefore was not submitted to a rigorous peer review process. *An Alternative Energy Strategy* falls into what can be described as 'grey literature', that is quasi-scientific reports that are not published in peer-reviewed journals but which are intended to reach a wider audience. However, while intended to be accessible to a wider audience, *An Alternative Energy Strategy* is not a work of popular science in the way that publications such as *Silent Spring* or Paul Ehrlich's *Population Bomb* are – it was never going to be a best seller. While there are technical terms, such as 'retrofitting', 'megawatt hours' and so on, these will be understood by most people, unlike some of the very technical jargon utilised in peer review articles intended for a specialist readership. Nonetheless, the use of what might be termed soft technical jargon alongside the use of numeric data contributes to the credibility of the argument.

The challenge for scientific texts intended for a wide readership, as Denise Tillery (2018: 44) points out, is 'how to make highly technical and specialised knowledge engaging and comprehensible for a large audience of non-specialists'. While the text of An *Alternative Energy Strategy* is comprehensible, I am not convinced that it is particularly engaging. The style is very different from Rachel Carson's *Silent Spring*, for example, which begins by drawing the reader into the argument with the short, poetic introduction 'A Fable for Tomorrow', which imagines a small rural town silenced by a blight of chemical pesticides. Carson's text is replete with metaphors and allusions such as 'the age of poison', 'elixirs of death' and so on, which have the potential to resonate with a wide readership and led to it becoming one of the best-selling environmental texts of all time.

There is a need to appeal to *pathos* – the emotions of the reader – rather than just making the rational argument. This is not to say that Carson does not make an appeal to reason (*logos*). However, Carson's main appeal is through *pathos*. While this strategy clearly resonated with a wide readership, Carson's style did lead to some criticism, particularly from within the scientific community, that it was too emotional, and therefore too subjective. *The Alternative Energy Strategy* opts for a different idiom and style, which is much drier, but nonetheless has its own persuasive aspects, which might be called the rhetoric of scientism. This utilises soft technological terms, frequently utilises simple quantitative data, tends to use a descriptive idiom, and makes extensive use of the passive voice and nominalisation.

When I asked Bob Todd who the intended readership for *An Alternative Energy Strategy* was, he primarily replied by focusing on the rationale to scale up CAT's ideas so that it would have a much wider resonance. In many ways, CAT, particularly in the early days, was an experimental laboratory for seeing what worked and what did not, and because of the scale of the experiment, it was frequently dismissed as not being more widely applicable. However, Bob Todd maintained that 'you could not write AT off on the basis that CAT was a small-scale experiment' (Bob Todd, interview December 2021).

The idea of scaling up from small-scale experiment to a wider application is consistent with the way in which ideas that are deemed to be successful in the laboratory are then developed for use in the wider world. However, Todd did indicate that it was intended to mail a copy of *An Alternative Energy Strategy* to every MP in the UK. Whether this was actually done or not is a moot point. CAT does suggest that 16 copies were sent Tony Benn's Ministry of Energy (CAT 2019: 5). Nonetheless, the desire to reach a wider, non-specialist audience does, to some extent, explain the style and framing of the *Alternative Energy Strategy*.

One of the ways by which we can determine who the intended readership is, and whom it is primarily intended to persuade is to look at how *An Alternative Energy Strategy* is framed. Matthew C. Nisbet and Todd P. Newman (2015: 325) suggest that frames are 'interpretative storylines that set a specific train of thought in motion communicating why an issue or a decision matter, who or what might be responsible, and which political actions should be considered over others'. Frames define the nature of the problem and therefore indicate the nature of the solution. Frames are organising devices that simplify complex issues and emphasise certain aspects of those issues. They are more effective if they draw upon pre-existing models of how the world works that are taken as common sense. As George Lakoff (2010: 72) suggests, communicating about the environment 'must make sense in terms of existing frames'.

The interpretive frame for *An Alternative Energy Strategy* is clearly economics. Framing environmental ideas in terms of economics is both a strength and a weakness. Policy makers need to consider financial aspects when making decisions. Consequently, the framing of the environment in economic terms makes sense as it refers to the common sense of governments as they primarily think in terms of budgets. However, while this framing can resonate with policy makers, it also tends to reduce environmental issues to a problem of resources. Framing the environment in economic terms also signifies that we are separate and distinct from the environment that surrounds us. Therefore, the solution is perceived as a straightforward accounting issue.

Frames are similar too, but different from metaphors. Arran Stibbe (2015: 64) suggests that 'metaphors use a frame from a specific, concrete and imaginable area of life to structure how a clearly distinct area of life is conceptualised'. As Stibbe indicates, there is a strong correlation between metaphor and framing. Both metaphors and frames are structured by a *source domain*, the area that is being drawn from, and the *target domain*, the subject of discussion. In this case, the frame target is the environment and the frame source is economics. Metaphors are more specific. For example, terms like capital and income are normally associated with economics (source) but when used to elucidate aspects of the environmental challenges (target) trigger a conception of the frame that equates addressing environmental challenges with solving economic issues. There are numerous metaphors, such as 'capital' and 'income', that trigger an economic frame of reference in *An Alternative Energy Strategy*.

> In an alternative approach to the formulation of an energy policy one first recognises that the earth's energy *capital* – the fuel reserves – is finite, and also that the energy *income* to the earth … is finite … The *capital* once used … cannot be reused … the energy *income* is inexhaustible.
>
> (3)

> *Reserves* and *income* of energy are both limited.
>
> (4)

> One also has to plan the gradual introduction of *income* energy – solar, wind, wave and tidal energy – as the use of fossil fuels is gradually reduced.
> (Todd and Alty 1977: 3–4. My emphasis)

In these quotes, energy is equated with the basic economic relationship between capital, income and expenditure. In this sense, economic terms are utilised as metaphors for energy resources. As George Lakoff and Mark Johnson (1980: 5) observe, 'the essence of a metaphor is understanding and experiencing one thing in terms of another'. Just as in economic terms we should live on income – an ongoing process – and not squander capital – a finite resource – the same applies to environmental resources. Fossil fuels are equated with capital, and if we continue to deplete capital, we will inevitably become destitute. On the other hand, renewable energy sources are equated with income and are 'abundant' and 'inexhaustible'. However, it is recognised that 'the perpetual flow of energy emanating from the sun, which manifests itself on earth as sunshine, wind, wave, tidal and biological energy arrives at a large but finite rate' (Todd and Alty 1977: 3). In other words, we should live within our (energy) means.

Later, the authors suggest:

> The adoption of the alternative energy strategy described would be the first step in the inevitable long-term adaptation to living *in balance with income energy sources.*
>
> (Todd and Alty 1977: 27)

This can also be considered as an economic metaphor – suggesting that we need to keep our energy books balanced. This is a matter of accounting, and disaster awaits all who fail to live within their means. Balance is a common trope in environmental discourses, which suggests that modern industrial society and consumer culture are out of balance with nature. Balance is perceived as being inherently good, and so we need to restore a more balanced relationship with nature.

Metaphors are not synonyms. We know that environmental concerns, although they overlap, are not the same as economic concerns. Nonetheless, as Lakoff and Johnson suggest, the metaphors that we use to communicate influence 'the way that we think, what we experience and what we do' (Lakoff and Johnson 1980: 3). While there are problems in framing environmental issues in economic terms, the metaphors and frames do have a greater potential to resonate with policy makers. As Nisbet and Newman (2015: 330) suggest, 'if a frame draws connections that are not relevant to something a segment of the public already values or understands, then the message is likely to be ignored or to lack personal significance'.

Peterson et al (2010: 114) have noted that there has been a comprehensive attempt to 'reframe ecologic concerns in economic terms to enhance environmental protection', which articulates environmental concerns in terms of the logics of 'neoliberal economic and political systems'. This discourse is referred to as ecosystems services, which suggests that ecosystems have an instrumental value, and therefore there is a direct benefit to humanity to protect the environment. Consequently, the framing of energy policy in terms of economics will resonate and therefore be more persuasive to policy makers than say framing energy policy in terms of aesthetics or nature.

Simon P. James (2015: 66) notes the ecosystem services discourse 'is not simply a rhetorical device: in many cases, the framework can be used to show that it really does make economic sense to protect nature'. Todd and his co-authors use economics to not only frame their argument and use metaphors such as capital and income, they also make a more direct economic argument. The environment is not an isolated issue and is also directly imbricated with economic concerns. For example, An *Alternative Energy Strategy* indicates that:

> It has been estimated that offshore wind energy systems would cost £440 per installed kilowatt with the result that they could provide energy at about 1.2p per kWh ... and will be economic if the price of oil to the power station exceeds £44 per ton ... The present world price of crude oil

is about £52 per ton. Given recent predictions, by the OECD and others, that the demand for oil will exceed supply by the mid 1980s ... The economic arguments for using wind energy will consequently get progressively stronger.

(Todd and Alty 1977: 9)

This direct economic reasoning bolsters the more metaphorically framed aspects of the argument.

Economic terms are not the only metaphors that can be identified in *An Alternative Energy Strategy*. An agrarian frame is also used. Alternative energy sources are said to be harvested.

In harvesting for our own purposes the mechanical energy of wind and wave or the heat energy of sunshine we merely redirect in its natural process of decline from mechanical energy to heat energy, and from hot to cold. We do not disturb the balance.

(Todd and Alty 1977: 4)

The three significant interrelated lexical choices in this short extract are 'harvesting', 'natural process' and 'balance'. Harvesting is an interesting metaphor to use in this context, and again is consistent with an ecosystem service frame. Generating power through renewable sources is equated with an agrarian, rather than an industrial practice. While cultivating and harvesting crops is perhaps not as close to nature as hunter-gathering, it is generally perceived as working with nature. Industry and, in particular, the extraction and use of fossil fuels are in contrast represented as exploiting nature. Harvesting signifies the bounteous aspects of agrarian practices that are cyclical and therefore in accord with natural processes. As the quote above indicates, *An Alternative Energy Strategy* equates harvesting with borrowing or redirecting natural processes. Renewable energy borrows power from nature rather than taking it, as in the extraction of fossil fuels. Once the mechanical energy of flowing water has been redirected through a turbine to utilise its power for our own purposes, the water continues to flow to the sea without harming the natural cycle.

The idea of nature and balance is commonplace in environmental rhetoric. A commonplace is a sort of argumentative shorthand, a condensation of a much more complex set of ideas and values that will be readily understood by the intended audience. The articulation of nature with balance becomes particularly significant with the development of ecology and the concept of ecosystems. The relationship between all the organisms and their environment in a particular ecosystem must be maintained in a balance for the mutual welfare of all life. Any disruption to the equilibrium of an ecosystem threatens all life. The idea of balance becomes even more significant with the development of James Lovelock's theory of Gaia, which perceives the whole planet as a self-regulating system. Disturbing the balance of nature therefore endangers not just a specific species of plant or animal; it is not simply a threat to a particular ecosystem but has the potential to imperil the planet. The authors indicate that harvesting the renewable natural resources of wind, wave and solar does not disrupt the balance of the Gaian system in the way that extractive industrial processes do. The balance is maintained and therefore the life of this planet, including humanity, is no longer threatened.

Zero Carbon Britain: An Alternative Energy Strategy

Many of the ideas and tropes that can be identified in *An Alternative Energy Strategy* are developed further in CAT's ongoing project ZCB. ZCB was directly inspired by *An Alternative Energy Strategy*. By the early 2000s, it became impossible for anyone involved in the environmental movement to ignore the urgency of addressing the climate emergency. It became clear that the CAT model proposed in *The Alternative Energy Strategy* from 1977 was not going to address the devastation of climate change. It was also realised that the situation was far more urgent than CAT had envisaged in the 1990s.

Consequently, a small group from CAT decided to revisit the *Alternative Energy Strategy*. Paul Allen, who is now ZCB Knowledge and Outreach Coordinator, was given half a day a week to work on the project. With the help of lecturers and students from the Graduate School of the Environment and consultation and various academics like Godfrey Boyle,[2] CAT produced the first report – *Zero Carbon Britain: An Alternative Energy Strategy*. This first ZCB Report proposes a pathway for Britain to achieve net zero carbon emissions within 20 years. This Report was published in 2007 with the aid of a small grant from the Esmée Fairbairn Foundation. Lembit Öpik, the local Liberal Democrat MP, attended the launch of the report and took several copies away with him. The Liberal Democrats announced their own policy initiative – *Zero Carbon Britain: Taking a Global Lead* at their party conference in 2007. Although it did not go as far as CAT's proposals, and did not formally acknowledge CAT's report, it does seem to have been influenced by ZCB. An article in *The Guardian* suggested that the Liberal Democrats' report 'was the most ambitious blueprint for climate change reform ever produced by a mainstream political party' (White 2007).

> We had launched a report and then a mainstream political party had a zero-carbon target. That is the first time that ever happened in British politics. So suddenly we were in this sort of policy influencing world.
>
> (Paul Allen, Oral History)

Allen also suggests that the Liberal Democrats' proposal created a demand for the CAT report that it otherwise might not have garnered, and that he and others involved in the project were increasingly invited to present on how to achieve Zero Carbon. In a Committee Discussion of the UK Parliament in 2008 on the future energy of Wales, CAT was directly mentioned by the Plaid Cymru MP Adam Price who stated:

> Obviously, the extent and composition of electricity generation is critical in terms of climate change. As Paul Allen of CAT said, in terms of renewable energy and latent potential, Wales could be the Saudi Arabia of Europe.
>
> (Price 2008, Column 59)

This clearly signifies that CAT had well and truly come in from the cold.

At the time of the first ZCB report, the Royal Commission on Environmental Pollution aspired to a 60% reduction in greenhouse gas emissions from the current levels by 2050 in the UK. CAT was the first organisation to argue for the imperative to achieve net zero. The ZCB report argued that the UK must not only aspire to net zero carbon emissions, but that this is an achievable goal. This was ahead of anyone publicly declaring the imperative to achieve net zero.

> CAT was able to lead the way because it could see the same writing on the wall as everybody else but was not trammelled by group or political pressures. Since then, many regional bodies have declared for net zero by 2030. So now the earlier studies, originally considered beyond the pale, are looking perfectly reasonable.
>
> (Peter, personal correspondence)

The ZCB Report

The first page of the first ZCB Report opens with a slightly out of focus photograph of Big Ben against a blue sky and viewed at a slight diagonal. The following page has a clear photograph of a footpath sign pointing in the direction of Whitehall. Below the image of the footpath sign is a quote attributed to John F. Kennedy.

> The problems of the world cannot possibly be solved by sceptics or cynics whose horizons are limited by the obvious realities. We need men to dream of things that never were.
>
> (cited in Helweg-Larsen and Bull 2007: 3)

The blurred image of Big Ben suggests that the government has lost focus on the issues that really matter. The representation of the fuzzy clock face of Big Ben signifies that the government is not only short-sighted but also seems oblivious to the fact that 'the clock is ticking' and there is an urgency to address the issues detailed in the report. The following image reinforced by the Kennedy quote suggests that the writers of the report are clear-sighted because their vision is not obscured by the limits of short-termism and scepticism. At the same time, the reference to the necessity to dream in the quotation alludes to the utopian aspect of ZCB. This utopian ideal and implicit critique of policy makers that is visually represented by these images and anchored by the quotation are emphasised in the short Executive Summary that follows.

This Executive Summary begins with the claim that ZCB is 'a radical vision of Britain's energy future'. Many of the current members perceive that ZCB continues the radical aspect of CAT's project.

> I think we are at the forefront. I think we encourage radical thinking. We encourage people to think differently, and we encourage them to visualise a different future.
>
> (Amanda, interview)

The perceived radical nature of the report is emphasised by a subheading 'Mapping the Unthinkable'. This reinforces the practical utopianism of CAT. The 'unthinkable' suggests a better way of being that has so far eluded the mainstream. The metaphor of a map signifies 'a journey from the present to a possible future … how to get from here to there' (Wright 2010: 26). Specifically, the journey is from an unsustainable present that threatens catastrophic climate change to a sustainable future. However, there is an urgency to the situation, and we need to move much more rapidly than is acknowledged by policy makers. The report unequivocally states that if the planet is to avoid catastrophic climate change, there is 'a degree of urgency that is lacking in official policy' (Helweg-Larsen and Bull 2007: 4).

Later in the report, it is suggested that '**zero**carbon**britain** is an intentionally bold title' (Helweg-Larsen and Bull 2007: 13). The use of the term 'bold' suggests that while courageous and enterprising, the proposals in the Report are not rash or foolhardy. To put this another way, the scenario for achieving net zero by 2027 is a radical proposal, but not so radical as to be perceived as either impossible or too extreme for the intended reader. The term 'bold' also has connotations of being strong and clear, and thereby signifies the strength and clarity of the argument. The boldness of the vision is reinforced using bold font for 'zero' and 'britain'.

In Arran Stibbe's terms, the ZCB project provides an alternative 'story-to-live-by'. Stibbe argues that the stories-we-live-by have several significant characteristics. Stories-that-we-live-by are shared across a culture, they shape how we perceive the world, they are often taken for granted as common sense, and they influence how we *act* in the world (Stibbe 2015: 6). Stibbe refers to the *Dark Mountain Manifesto*, in which the authors argue that the dominant story of the contemporary world is 'the centrality of humanity and humanity's 'mastery over "nature" to which we no longer belong'. They further claim that that 'what makes this story so dangerous is that, for the most part, we have forgotten it is a story' (Kingsnorth and Hine 2009). This conflation of perception with reality poses a serious challenge for anyone who proposes an alternative story. To be persuasive, alternative stories must be couched in a way that do not appear to conflict with an assumed reality. Direct contradiction of a taken-for-granted perception of the world is likely to alienate the reader. The radical agenda of alternative stories to live by must be articulated in a recognisable mode. Consequently, the ZCB report is presented in a very conventional mode in order to avoid being dismissed as non-sensical heresy. ZCB adopts three main strategies to mitigate its 'radical agenda' it refers to mainstream institutions and recognised experts, it frames its argument in economic terms and it utilises a number of conventions used in the language of science.

Reference to mainstream institutions and experts reinforces the credibility of the argument. Most noticeably, the report begins with a short Foreword by Sir John Houghton who was a former chair of the IPCC and a former Director of the UK Meteorological Office. Sir John Houghton commends the authors of this report for 'their imagination (coupled with realism), their integrated view and their sense of urgency' (cited in Helweg-Larsen and Bull 2007: i). This is clearly a recognition of the practical utopianism of the report. The report also refers to official reports such as the Stern Review on the Economics of Climate Change.[3] For example the report cites Nicholas Stern's observation that 'climate change threatens the basic elements of life for people around the world – access to food, production, health, and use of land and the environment' (in Helweg-Larsen and Bull 2007: 14). Here are two stalwarts of science and economics that appear to support the basic premise of the ZCB position. This signifies that CAT is in accord with recognised authoritative voices, and therefore, the alternative story of ZCB cannot simply be dismissed as implausible.

There is a clear tension here. Mapping the unthinkable suggests that CAT is situated, if not outside then at least, on the margins of mainstream acceptability. CAT is not constrained by norms and protocols of formally established institutions, yet at the same time seeks legitimacy from the establishment. It could be argued that CAT wants its cake and to eat it – to claim legitimation by established institutions and at the same time challenge those same institutions. However, I would argue that this ambiguity is precisely CAT's strength. CAT has from its very beginning had an ambiguous relationship with the establishment. In some ways, this reflects Gerard Morgan-Grenville's ambivalent status of being both in and outside the establishment. CAT's ethos is best captured by the suggestion that at its best it is 'radical but not too radical'. If CAT is perceived as being 'too radical', it can be dismissed as being a small group of disenfranchised 'hippies on the hill'. On the other hand, if it is not radical enough, CAT is in danger of being accused of being ineffective as they have compromised too much with establishment values. This is a fine balance and CAT has not always got right, but over the years, it does seem to have developed more finesse in this ideological juggling.

The ZCB report, like its predecessor *An Alternative Energy Strategy*, frames much of its argument in economic terms. However, economics is reimagined in terms of carbon, rather than finance.

> The proposal is a shift from the traditional financial economy of today into an economy sensitive to carbon. The core dynamic and drivers are

different. Maximising profit will always drive decisions. But in tomorrow's economy, decisions on every level – personal, business and governmental – will be driven by the imperative to conserve valuable carbon permits and to pursue cheaper, zero-carbon enterprises.

(Helweg-Larsen and Bull 2007: 14)

In other words, carbon trading is not considered simply as a sort of parallel economy but should also inform decisions in the financial economy.

The ZCB Report is very similar in style to *An Alternative Energy Strategy* and adopts many of the conventions of grey literature in that it is not written for the popular market or for a specialist scientific community. It is intended to be taken seriously by a readership who might be resistant to the alternative story advocated by the content of the Report, and consequently, it utilises a familiar style. This strategy of new wine in old wine skins is an attempt to make the alternative story if not palatable at least credible to a potentially antagonistic audience.

The use of soft-technical terminology also reinforces the credibility of the report. For example, terms like CO_2, greenhouse gas emissions and ecosystems are used throughout the report. Harder technical terms are also occasionally used, and these are accompanied by short explanations. This use of more specialised terminology coupled with clear explanations reinforces the credibility and expertise of the authors. For example, in the broad overview of climate change, the authors refer to 'the albedo effect', which is a term that is not widely known. Consequently, there is a text box that indicates that 'albedo refers to the reflectivity of the planet's surface' and then explains the relevance of albedo to climate change. This text box also refers to various untried technical solutions, such as cloud seeding, which propose to manipulate albedo to mitigate global warming. The authors are emphatic that these untested large-scale technological fixes are highly uncertain. (Helweg-Larsen and Bull 2007: 21).

The use of technical and scientific terminology is augmented by utilising the conventions of scientific language. The report predominantly utilises a passive voice and makes extensive use of nominalisation, which 'possesses features of condensed information, concise expression, compact structure and strong logic. Hence nominalization is often used in scientific, legal and political style, because these styles are comparatively formal' (Lei and Yi 2019: 98). The report is presented in a formal style that is consistent with conventional scientific writing and therefore signifies a neutral, rational and objective account that has the potential to mitigate the more radical aspects of the argument.

Nominalisation also erases specific agents who might otherwise be held accountable for specific actions. Linguists normally evaluate the use of the nominalisation negatively as it hides the actors who are responsible for destructive behaviour. Stibbe (2015: 148) suggests that 'if the key actors responsible for environmental destruction are systematically erased from environmental discourse then the danger is that solutions are sought at the wrong level'. However, this use of nominalisation and concomitant erasure of agents is not always a negative. I agree with Tim Jensen's observation that the climate crisis is 'collectively created, without question, but unevenly so, with some actors who are far more culpable, yet who feel less responsible' (Jensen 2019: 35). However, apportioning blame to agents who do not feel personally culpable is not an effective rhetorical strategy. Blame will provoke defensiveness, which will not encourage the responses required to address the climate emergency. It is better to frame the climate emergency in more neutral terms as an objective reality, which it is imperative that we all address, and then provide some plausible solutions to the challenges, as the ZCB Report does.

The plausibility and objectivity of the argument of the ZCB strategy is reinforced by the extensive use of quantitative data throughout the report. It is generally thought that quantitative data are essentially objective – simply a matter of counting some independent aspect of the world. However, it is increasingly recognised that numeric data are rhetorical. James Wynn and G. Mitchell Reyes (2021: 21) suggest that 'mathematics, because of its perceived transparency and objectivity' can 'replace personal trust as a persuasive force'. In other words, statistical information can potentially stand in for personal credibility (*ethos*). Despite a growing scepticism about quantitative data and the uses and abuses of statistics made by politicians to bolster their agenda, there is still a faith in numbers. Theodore Porter in his classic work *Trust in Numbers* observes there is a link between quantification and objectivity. He suggests 'quantification is well suited for communication that goes beyond the boundaries of locality and community. A highly disciplined discourse helps to produce knowledge independent of the particular people who make it' (Porter 1996: ix).

This quantitative aspect of the report is visually reinforced by graphs and diagrams, which have their own rhetorical character. Graphs and diagrams are a technique to visually represent the quantitative data with the aim of aiding the reader's understanding. 'An effective use of graphs and diagrams provides the viewer the greatest number of ideas in the shortest time with the least ink in the smallest space' (Tufte 1983: 51). Advances in the technology of publishing have facilitated the inclusion of graphics into reports. 'Computer-generated displays initiated a critical shift in the rhetorical process by substantially altering reader expectations' (Kostelnick 2007: 284). Consequently, there is now almost an imperative to include charts and diagrams to enhance the credibility of quasi-scientific reports. Charles Kostelnick (2007: 284) observes that, 'a well-executed display will enhance its clarity and its ethos'.

The diagrams and charts are much more detailed and sophisticated than the graphs used in the 1977 *An Alternative Energy Strategy*. However, while graphs, diagrams and charts are intended to communicate complex ideas with clarity, as Kostelnick notes, clarity is always contingent upon the interpretive aptitude of the viewer. 'A highly technical chart that appears in a journal for high energy physicists may be perfectly clear to that audience, but that same chart may be inscrutable to lay readers' (Kostelnick 2007: 283). Overall, the diagrams and charts used in the report are clear and simple and elucidate the argument made in the text using genres of charts, such as bar, pie and line, that will be familiar to the reader. These graphs visually display the overall argument and rhetorically contribute to the credibility of the report. For example, there is a colourful graph that displays a month-by-month projection of electricity demand and potential supply from renewable sources, such as offshore wind and photovoltaics, for the UK. The graph visually reinforces the claim that Britain can be reliant on renewable energy sources.

The central argument of the report is articulated in the dual processes of 'power down' and 'power up'. Powering down refers to the need to reduce overall energy demand. The report suggests that 'Britain is energy obese: far more is used than is actually required to deliver well-being' (Helweg-Larsen and Bull 2007: 7). This frames the argument in terms of health and wellbeing. While there are many complex reasons for obesity, there is a common perception that the prime cause is overeating, particularly of poor-quality foods. The implication is that as a nation we are overindulging in poor-quality energy that directly impacts on the health and wellbeing of the population and the planet. The metaphor of obesity indirectly counters the perception that consuming less entails a drop in the standard of living. This is more directly addressed when the authors state that 'powering down does *not* mean deprivation, or a return to hardships of the past'. However, just as the obese need to alter their lifestyle habits in relation to food consumption, we also must adapt our lifestyles to healthier, less carbon-dependent ways of being (Helweg-Larsen and Bull 2007: 7).

Powering up refers to the increasing use of renewable energy sources. Powering down using the various strategies outlined in the report, such as making housing more energy efficient, dramatically reduces energy demands in the UK. This lower energy demand can then be met by increasing (powering up) use of renewable sources. The end of the ZCB Report clearly proclaims in large green font: 'We have more than enough renewable sources of energy to power up. The technical wind and wave resource off the coast of Britain is massive …'. The large green font reinforces the claim that the goal to supply the UK's energy demand through renewables is a plausible alternative and of course the colour green is now widely associated with environmentally friendly attitudes.

Throughout the report, the authors attempt to balance this idealistic vision of a zero carbon future with a wide range of devices to emphasise the viability and achievability of the suggested alternatives. The ZCB report provides 'utopian ideals that are grounded in the real potentials of humanity, utopian destinations that have accessible waystations, utopian designs of institutions that can inform our practical task of navigating a world of imperfect conditions' (Wright 2010: 6). Restating Wright's characterisation of a real utopia, ZCB is a vision of a zero carbon ideal that is grounded in the real potential of humanity, that has viable goals along the path and plausible transformations of social institutions to ensure that net ZCB is achievable.

Developing the ZCB Project

In 2008, CAT returned to the Esmée Fairbairn Foundation with a proposal to update the ZCB report in the context of the financial crisis. The Esmée Fairbairn Foundation, which not only funds proposals on environmental issues but also finances projects that are concerned with social justice, suggested that CAT should work with the New Economics Foundation (NEF), whose mission is 'to transform the economy so it works for people and the planet' (NEF 2021). NEF had published a report from the Green New Deal Group, which envisages 'a new economy shaped to prevent a climate breakdown and transform a failed financial system' (The Green New Deal Group 2021). This collaboration produced the second report *Zero Carbon Britain 2030: A New Energy Strategy*, which was published in 2010. While CAT has never worked in isolation, this does mark a more formal collaborative approach and further expanded CAT's networks of influence.

This second report was presented to the All-Party Parliamentary Climate Change Group. Paul Allen recalls personally giving a copy of the report to Chris Huhne, who was the Secretary of State for Energy and Climate Change in the Coalition Government at the time.

> It was very exciting because we were in Parliament with the decision makers ... Suddenly we felt like we were getting a radical message right home to a lot of big thinkers.
>
> (Paul Allen, Oral History)

Even in 2014, the fifth Report from the IPCC only advocated 'near zero emissions of carbon dioxide and other long-lived greenhouse gases by the end of the century' (IPCC 2014). In 2014, a year before the 21st Conference of Parties (COP21) held in Paris, Jim Yong Kim, the President of the World Bank, gave a speech in which he asserted that net zero emissions of greenhouse gases must be achieved before the end of the century.[4] COP21 produced what has become known as the Paris Climate Accords, in which the signatories agreed that greenhouse gas emissions must be reduced as rapidly as possible, and that net zero must be achieved by the middle of the next century. Sweden was the first country to enshrine this in law in 2017, but it was not until 2019 that the UK government became the first major economy to legislate to achieve net zero greenhouse gas emissions by 2050. Although it was published in 2007, CAT's *Zero Carbon Britain: An Alternative Energy Strategy* was more ambitious as it suggested reducing carbon emissions to zero within 20 years. This was slightly amended in the second *Zero Carbon Britain Report: A New Energy Strategy* published in 2010 to 'reduce our greenhouse gas emissions to zero as fast as possible, in this report we adopt 2030 as our target year' (CAT 2010: 2).

CAT has produced a further three reports that have been based on new information on the climate emergency.[5] The latest 2019 Report states:

Building on the groundwork laid by the Zero Carbon Britain Project over the last 12 years, we incorporate the latest developments in science and technology in the key areas of balancing highly variable energy supply and demand, and the nutritional implications of a low carbon diet.

(CAT 2019: 4)

Here CAT indicates its credibility by emphasising that not only does it have a relatively long history of investigating the climate emergency, but that it is also aware of all the latest developments in climate science.

These updated reports follow a very similar format to that of the earlier reports – with extensive use of charts, nominalisation and the passive voice. In addition to the five Zero Carbon Britain Reports, CAT has published two supplementary reports, *People, Plate and Planet* (2014) and *Raising Ambition: Zero Carbon from Across the Globe* (2018). *People, Plate and Planet*, as the title suggests, is focused on diet and food production. While diet and land use had been covered in the earlier ZCB reports, *People, Plate and Planet* looks at specific issues, such as the environmental impact of reducing food imports, in more detail. This report also emphasises that a low carbon diet tends to be better nutritionally and therefore potentially can improve individual health and wellbeing. There is a consistent emphasis in the ZCB discourse that, while reducing carbon use is clearly critical if we are to avert environmental devastation, there are also many other spin-off benefits to be gained by this.

While the ZCB project is primarily focused on how the UK can decarbonise, it also acknowledges that the climate emergency is a global issue that cannot be solved by Britain alone. *Raising Ambition: Zero Carbon from Across the Globe* includes case studies from around the world 'From Tanzania to Los Angeles, South Asia to the Baltic' that have modelled 'net zero, deep decarbonisation, and up to 100% renewable energy' (Allen and Bottoms 2018: 7). This emphasises that CAT is part of a global network of groups and organisations that not only accept the reality of the climate emergency but also recognise the imperative to identify practical solutions.

While there is much to admire, and while I agree with the ecosophy of the ZCB Reports, I am not convinced that they are as inspiring as they could be. This is in part due the style in which they are written and the fact that CAT is not particularly sure of who the readers of these reports are. When I asked one of the prime leads in the Zero Carbon Britain project who the intended readers for the reports are, he suggested:

I don't think of aiming it at something. Zero Carbon Britain is not a rifle, it is a tool. It is something that you offer rather than targeting people. It is a conversation starter.

(Paul, interview January 2021)

Overall, there is too much detail in all the ZCB Reports for the lay reader and they are too general for the specialist.

The challenge for CAT and all environmental groups is that for the climate crisis to be effectively addressed the audience has to include everybody. However, the audience is never homogenous and comes with diverse presuppositions, knowledge and perspectives. Not everyone who encounters CAT is going to be persuaded. However, CAT is a significant voice in an environmental network that is amplified by other concerned voices and in turn helps augment these other voices, which together creates a more plausible and persuasive discourse.

Making It Happen

Zero Carbon is now well established on the political agenda of most major economies. The UK Government produced a report *Net Zero Strategy: Build Back Greener* in October 2021.[6] About 300 local councils in the UK have declared a climate emergency and have pledged to reduce their carbon emissions. Many businesses, from large multinationals to small local companies, have committed to decarbonising. The challenge, at all levels, is how to actually decarbonise our activities. While the ZCB Reports provide a useful overview of how we can achieve net zero by powering up our use of renewables and powering down our overall energy consumption, they do not provide much detail of how we might implement these changes.

Consequently, CAT has expanded the ZCB project in a number of different directions to help enable local governments, businesses and individuals implement decarbonisation. Perhaps, the most important of these new ZCB initiatives has been the founding of the ZCB Innovation Lab in 2020. Innovation labs have become increasingly popular in the last decade. Adam Wellstead et al suggest that they 'share three distinct features: the use of design-thinking methodology; a focus on innovation; and a user-centric approach'. Innovation labs 'create a collaborative space to enable participants with varied skill sets to reach a common understanding of a policy challenge' (Wellstead, Gofen and Carter 2021: 194). Using this approach, CAT's ZCB Innovation Lab aspires 'to help communities, local authorities and policymakers to create Zero Carbon Action Plans, and to provide support for the development of innovative solutions' (CAT 2022).

The policy challenge for councils, businesses and governments is how to decarbonise once they have made the commitment to net zero. As Anna Bullen observes, in a webinar on the importance of co-creation in addressing the climate emergency, the challenge of decarbonising is a wicked problem (Bullen 2021).[7] The climate emergency is a wicked problem because there is no single cause and no clearly identifiable solution. Wicked problems can 'engender a high level of conflict among the stakeholders as there is no agreement on the problem or its solution' (Roberts 2000: 1). It is well established that 'dialogic, two-way forms of communication are more conducive to fostering change than one-way information delivery' (Whitmarsh, O'Neill and Lorenzoni 2015: xvii). Bullen cogently argues in her webinar that imposing solutions from above is not effectively going to address the climate emergency and co-creation is the only way to identify and implement solutions. Consequently, the ZCB

Innovation Lab creates a safe space for stakeholders at all levels of a council or organisation to recognise the challenges and to propose feasible solutions. This is not CAT providing the solutions but facilitating stakeholders in co-designing the solutions that are apposite to their own particular contexts.

For example, CAT worked with the 10 Staffordshire councils – County, City, Borough and District – to help them identify and implement solutions to reduce carbon emissions, after the County Council declared a climate emergency in 2019. CAT started with a survey to understand where the councils were in terms of their journey to net zero and then, in collaboration with the councils, determined a goal to 'work better together to address the climate and ecological crisis'. CAT then organised a series of workshops over a ten-month period that brought together 32 participants, from Project Officers to elected members. This produced a series of recommendations that would help the councils work more effectively together to achieve net zero (Bullen 2021).

> The first workshop was difficult for them, they had to look at things in a very different way to how they would normally look at things. I think they really thought they were going to turn up and we were just going to hand over the solutions.
>
> (Anna, interview July 2021)

In some ways, this collaborative, co-creative approach to addressing the climate emergency harkens back and extends the egalitarian and consensus decision-making ethos of the early days.

> I really wanted the innovation lab to grow from CAT to recognize that CAT was and is its own innovation lab, you know.
>
> (Anna, interview July 2021)

A participant in these workshops indicated that they were organised very professionally, but perhaps like the ZCB Reports were overly ambitious.

CAT now provides a whole series of training workshops and events focused on various aspects of achieving zero carbon. For example, Zero Carbon Britain Live Online – a two-day interactive online course 'offers the hard data and confidence required for visualising a future where we have risen to the demands of climate science. It helps to reduce fear and misunderstandings and open new, positive, solutions-focused conversations' (CAT 2022). CAT has also compiled a Resource Hub, with a plethora of links to various other organisations, reports and training programmes for decarbonisation.

The Zero Carbon Britain project has now become the central focus of CAT's agenda alongside the Graduate School.

> Everything can now be set in the context of climate change and that unifies everything because it's all about cutting carbon emissions.
>
> (John C, interview October 2020)

Zero Carbon is now well integrated into wider social, political and media discourses. *The Sun*, the UKs best-selling newspaper by far, suggested in a headline in March 2021: 'ZERO PLANS. Ministers blasted for "still having no plan" for how to get UK to zero carbon emissions by 2050' (Clark 2021). Consequently, the idea of zero carbon will have some familiarity for most people. There is now a ZCB Trail on site that 'lets you explore this [ZCB] research in bite-size digestible chunks, helping you think about the future in a different way' (CAT 2019: 8). Clearly, not everyone will join a webinar, enrol on a GSE course or read the reports – but the information on how we can plausibly achieve net zero might inspire some to seek more in-depth information and even consider how they might reduce their own carbon footprint.

Notes

1 At the time of publication of *An Alternative Energy Strategy*, there was little understanding of the impact that burning fossil fuels has on GHG emissions and climate change. Now of course, there would be absolutely no consideration of any use of coal to generate power.
2 The founder and editor of *Undercurrents*, co-author with Peter Harper of *Radical Technology* and the creator of the Alternative Energy Group at the Open University.
3 A report, written by Nicholas Stern, a reputed economist, about the economics of climate change. It was commissioned by the UK Chancellor of the Exchequer and the Prime Minister and published in 2006.
4 The full speech can be found at www.worldbank.org/en/news/speech/2014/12/08/transforming-the-economy-to-achieve-zero-net-emissions
5 *Zero Carbon Britain: 2030 A New Energy Strategy* (2010); *Zero Carbon Britain: Rethinking the Future* (2013); *Zero Carbon Britain: Making It Happen* (2017); *Zero Carbon Britain: Rising to the Climate Emergency* (2019). All of these reports are available as PDFs from CAT's website https://cat.org.uk/info-resources/zero-carbon-britain/research-reports
6 Available at www.gov.uk/government/publications/net-zero-strategy
7 Available at https://cat.org.uk/past-webinars

References

Allen, P. & Bottoms, I. (2018). *Raising Ambition: Zero Carbon Scenarios from Across the World*. Machynlleth: CAT Publications. Available at: https://cat.org.uk/info- resources/zero-carbon-britain/research-reports
Başçi, S. & Hassan, N. (2020). Using Numbers to Persuade: Hidden Rhetoric of Statistics. *International Econometric Review* 12(1), 75–97. doi: 10.33818/ier.747554
Brighton, P. & Foy, D. (2007). *News Values*. London: Sage.
Buell, F. (2004). *From Apocalypse to Way of Life: American Crisis in the American Century*. London: Routledge.
Bullen, A. (2021). Addressing the Climate Emergency Together: The Importance of Co- Creative Practice. Available at: https://cat.org.uk/past-webinars (recorded 9 November 2021). Centre for Alternative Technology (CAT). (1995). *Official Guide Book*. Machynlleth: CAT Publications.

Centre for Alternative Technology (CAT). (2010). *Zero Carbon Britain: 2030 A New Energy Strategy.* Machynlleth: CAT Publications. Available at: https://cat.org.uk/info- resources/zero-carbon-britain/research-reports

Centre for Alternative Technology (CAT). (2019). *Zero Carbon Britain: Rising to the Climate Emergency.* Machynlleth: CAT Publications. Available at: https://cat.org.uk/info- resources/zero-carbon-britain/research-reports

Centre for Alternative Technology (CAT). (2022). *New Hub and Innovation Lab To Share Zero Carbon Solutions.* Available at: https://cat.org.uk/new-hub-and-innovation-lab-to-share-zero-carbon-solutions (Accessed 20 April 2022).

Clark, N. (2021). ZERO PLANS. Ministers Blasted for "Still Having No Plan" for How to Get UK to Zero Carbon Emissions by 2050. *The Sun* (5 March 2021) Available at: www.thesun.co.uk/news/14240611/ministers-zero-carbon-emissions-2050 (Accessed 17 February 2022).

Djerf-Pierre, M. (2012). When Attention Drives Attention: Issue Dynamics in Environmental Reporting Over Five Decades. *European Journal of Communication* 27(3), 291–304. doi: 10.1177/0267323112450820

Dryzek, J.S. (2005). *The Politics of the Earth: Environmental Discourses.* Oxford: Oxford University Press.

Green New Deal Group. (2021). About the Group. Available at: https://greennewdealgroup.org/about-the-group (Accessed 30 November 2021).

Hansen, A. (2019). *Environment, Media and Communication.* London: Routledge.

Helweg-Larsen, T. & Bull, J. (2007). *Zero Carbon Britain: An Alternative Energy Strategy.* Machynlleth: CAT Publications. Available at: https://cat.org.uk/info-resources/zero-carbon-britain/research-reports

Intergovernmental Panel on Climate Change (IPCC). (2014). *AR5 Synthesis Report: Climate Change 2014* Available at: www.ipcc.ch/report/ar5/syr (Accessed 24 November 2021).

Intergovernmental Panel on Climate Change (IPCC). (2021). *History of the IPCC.* Available at: www.ipcc.ch/about/history (Accessed 23 November 2021).

James, S.P. (2015). *Environmental Philosophy: An Introduction.* Cambridge: Polity Press.

Jensen, T. (2019). *Ecologies of Guilt in Environmental Rhetorics.* Cham, Switzerland: Springer Nature.

Kearns, G. (1991). Green Idealism Versus Tory Pragmatism. *Journal of Biogeography* 18(4), 363–70. doi: 10.2307/2845478

Kim, J.Y. (2014). *Turn Down the Heat: Confronting the New Climate Normal.* Washington: The World Bank Group Available at: www.worldbank.org/en/topic/climatechange/publication/turn-down-the-heat (Accessed 24 November 2021).

Kingsnorth, P. & Hine, D. (2009). *The Dark Mountain Manifesto.* Available at: https://dark- mountain.net/about/manifesto (Accessed 1 February 2022).

Kostelnick, C. (2007). The Visual Rhetoric of Data Displays: The Conundrum of Clarity. *IEEE Transactions on Professional Communication* 50(4), 288–94. doi: 10.1109/TPC.2007.908725

Lakoff, G. (2010). Why It Matters How We Frame the Environment. *Praxis Forum* 4(1), 70–81. doi: 10.1080/17524030903529749

Lakoff, G. & Johnson, M. (1980). *Metaphors We Live By.* Chicago, IL: University of Chicago Press.

Lei, Y. & Yi, S. (2019). Researches on Application of Nominalization in Different Discourses. *International Journal of Research: Granthaalayah* 7(4), 97–105. doi: 10.29121/granthalayahvy.i4.2019.879

Lowery, S.A. & DeFleur, M.L. (1995). *Milestones in Mass Communication Research.* London: Longman.

New Economics Foundation (NEF). (2021). Our Work. Available at: https://newec onomics.org/about/our-work (Accessed 30 November 2021).

Nisbet, M.C. & Newman, T.P. (2015). Framing, the Media and Environmental Communication. In A. Hansen & R. Cox (Eds.), *The Routledge Handbook of Environment and Communication.* London: Routledge, 325–338.

Peterson, M.J., Hall, D.M., Feldpausch-Parker, A.M., & Peterson, T.R. (2010). Obscuring Ecosystem Function with Application of the Ecosystem Services Concept. *Conservation Biology* 24(1), 113–19. doi: 10.1111/j.l523-1739.200901305.x

Porter, T.M. (1996). *Trust in Numbers: The Pursuit of Objectivity in Science and Public Life.* Princeton, NJ: Princeton University Press.

Price, A. (2008). Future Energy of Wales. *HC Hansard*, Columns 58–9. Available at: https://publications.parliament.uk/pa/cm200708/cmgeneral/welshg/080618/pm/80618s02.htm (Accessed 30 November 2021).

Roberts, N. (2000). Wicked Problems and Network Approaches to Resolution. *International Public Management Review* 1(1), 1–19. Available at: https://journals.sfu.ca/ipmr/index.php/ipmr/issue/view/1

Stibbe, A. (2015). *Ecolinguistics: Language, Ecology and the Stories We Live By.* London: Routledge.

Tillery, D. (2018). *Commonplaces of Scientific Evidence in Environmental Discourses.* London: Routledge.

Todd, R. & Alty, C. (Eds.) (1977). *An Alternative Energy Strategy for the United Kingdom.* Machynlleth: National Centre for Alternative Technology.

Tufte, E.R. (1983). *The Visual Display of Quantitative Data.* Cheshire: Graphics Press.

Wellstead, A.M., Gofen, A., & Carter, A. (2021). Policy Innovation Lab Scholarship: Past, Present, and Future – Introduction to the Special Issue on Policy Innovation Labs. *Policy Design and Practice* 4(2), 193–211. doi: 10.1080/25741292.2021.1940700

White, M. (2007). Lib Dems See Zero-Carbon Britain Setting the Global Green Agenda. *The Guardian* (29 August) Available at: www.theguardian.com/politics/2007/aug/29/uk.greenpolitics (Accessed 30 November 2021).

Whitmarsh, L., O'Neill, S., & Lorenzoni, I. (2015). *Engaging the Public with Climate Change: Behaviour Change and Communication.* London: Earthscan from Routledge.

Wright, E.O. (2010) *Envisioning Real Utopias.* London: Verso.

Wynn, J., & Reyes, G.M. (2021). From Division to Multiplication Uncovering the Relationship Between Mathematics and Rhetoric Through Transdisciplinary Scholarship. In J. Wynn & G.M Reyes (Eds.), *Arguing With Numbers: The Intersections of Rhetoric and Mathematics.* University Park, PA: The Pennsylvania State University Press, 11–32.

Conclusion

The Centre for Alternative Technology (CAT) might seem to be an obscure environmental educational charity located in remote rural Wales. Although not as visible as campaigning groups such as Extinction Rebellion, CAT has had an impact that belies this apparent obscurity. This impact can be best understood in terms of accumulation – historical and pedagogical. Historically, CAT has accumulated a reputation that has enhanced its credibility over time. In part, this is because it has survived and adapted to the changing context, and in part because it has developed an expanding network of support. Pedagogically, CAT, as an educational charity, has always adopted a cumulative approach. The visitor centre, as Roger Kelly CAT's third director suggested, is intended to be the first rung on the ladder of environmental awareness. It is hoped that this will then inspire people to move further up the ladder. CAT can then offer, to those that are inspired, more in-depth information through its information services and educational courses.

CAT has had a relatively long history in terms of the environmental movement. It was founded in 1973 at a time when there was a growing awareness of the devastating impact of human activity on the environment. Friends of the Earth was founded in 1969, Greenpeace began in 1971, the first Earth Day was held in 1970, the UK's first green political party – the PEOPLE Party (renamed in 1975 as the Ecology Party) – was founded in 1972 and the *Ecologist* magazine published *A Blueprint for Survival* in 1972. This period was the beginning of what Joachim Radkau (2014) has called the Age of Ecology and was a particular rhetorical moment in time. In rhetorics, the term *kairos* which means time, has 'connotations of opportune moment, right measure, and appropriateness' (Carter 1988: 105). There was a receptive audience for CAT's message in the 1970s. However, while there was sufficient interest to sustain CAT, environmentalism was primarily a niche concern. CAT's main challenge was to extend its appeal beyond the already converted. Audiences are inevitably diverse and what is persuasive for some people will not necessarily resonate with others. Furthermore, as the context changes, what was opportune, of right measure and appropriate, might cease to be so. In other words, *kairos* is not a fixed point in time and varies according to the perspective of the audience. Consequently, the mode of address and the way in which an argument

DOI: 10.4324/9781003207702-7

is framed must be adaptable to be persuasive to as many people as possible at different times.

In order to address the environmental devastation caused by human activity, it is necessary to persuade the uninterested and the sceptical as well as the receptive audience. CAT is in a unique position. Although in a relatively remote rural area, CAT's location was serendipitous. The beautiful Dyfi Valley is on the southern edge of the Snowdonia National Park and in close proximity to the sea, and consequently the area is a popular tourist destination. A visit to CAT by tourists in the area is often included as one of their holiday activities. While there are mixed responses to CAT by the curious day visitor – some are indifferent, some bewildered and some even hostile – there is plenty of anecdotal evidence that for others the site was inspiring and stimulated an interest in environmental issues. For example, in CAT's latest annual report, a quote from a visitor states: 'My dad used to bring me in the 70s when I was little. I have taken my own children and always find inspiration and hope here' (cited in CAT 2022a). While this might be a very small pebble in a very large pond, these small ripples have a cumulative effect, which both reinforces and is reinforced by other voices in the environmental movement.

Over time, the mutually reinforcing voices of the environmental movement, coupled with growing scientific evidence about the human impact on the environment, amplified these small ripples of concern, which then became more widely accepted. By the early 1990s, it was no longer possible for politicians to ignore the plight of the planet, and concern about the environment was beginning to attract growing attention in the media and popular culture. Peter Rawcliffe (2000: 65) observes:

> The 10 years from 1985 to 1995 were a period of dramatic change in the environmental politics of Britain … The green tide which swept through the country was driven by the perception that the environment had evolved into a legitimate, high profile and enduring political and policy issue.

As environmental concern moved from a niche issue to being more widely incorporated into mainstream discourse, the mode of address has had to be adapted to ensure that it continues to be opportune, of right measure and appropriate. The major challenge for CAT, and other environmental groups, was the development of the discourse of ecological modernisation, which involved 'reconceptualising the relationship between the environment and the economy' (Young 2000: 2). In this discourse, economic growth and environmental protection are no longer perceived as incompatible.

What Stephen C. Young (2000) has called 'ecological modernisation' is a mixed blessing. On the one hand, it has meant that the message of the environmental impact of human activity is now widely accepted. In some ways, this can be seen as a major achievement of the environmental movement's cumulative voice and has given environmental NGOs greater credibility and an enhanced role in debates and discussions in the public arena. On the other hand, ecological modernisation suggests that we can have our 'environmental cake and eat it'. This is incredibly alluring as it suggests that the environmental crises can be solved without any major systemic changes to the capitalist system or individual lifestyle choices. It also enables policy makers to marginalise more radical and oppositional voices, as governments and businesses can now claim that they are actively addressing environmental concerns. This poses a major challenge for environmental groups, such as CAT, who believe that the planet cannot sustain continued economic growth and argue that radical systemic and lifestyle changes are needed to prevent humanity and the earth hurtling towards major environmental catastrophe.

To be persuasive, CAT, in the context of ecological modernisation, has had to adapt its rhetorical strategy. CAT has, like many environmental NGOs, become more policy orientated – particularly working with the Welsh Government, local councils and businesses who have pledged to de-carbonise their activities. CAT is well placed to do this for several reasons. CAT's focus has always been on identifying practical solutions. The challenge for businesses and policy makers, who are under increasing pressure to reduce their carbon emissions, are in dire need of practical advice. As CAT has survived for nearly 50 years, this cumulative history has contributed to CAT's credibility. CAT can no longer be simply dismissed as 'the hippies on the hill'. Finally, CAT has had considerable experience of knowing how far to push the envelope of acceptability – to appear radical but not too radical. It has made its radical message of decarbonisation more palatable by utilising the conventions of scientific discourse and framing much of the argument in economic terms.

One aspect of the emergence of ecological modernisation has been the development of the hardware of AT, and the concomitant rise of a 'green sector' that has generated employment and economic opportunities. No one at CAT, when I was there in the 1980s, envisaged the advances in renewable energy technology – the development of megawatt wind turbines, cost-effective solar panels and so on. The job opportunities for skilled engineers, builders, architects (and even caterers) who were concerned about sustainability were very limited. In its early years, CAT was one of the few places where it was possible to work with renewable energy. However, with the development of offshore and onshore wind farms, commercial and domestic use of PVs, increasing interest in heat pumps, escalating demand for sustainable building design and so on, there is a growing demand for skilled labour in the green sector. CAT has created a niche through its various MSc programmes on sustainability. Many of the graduates from the Graduate School of the Environment (GSE) have not only started small businesses in various aspects of sustainability or found employment in the green economy but have also become ambassadors for CAT.

In this new context of ecological modernisation, Schumacher's philosophy 'small is beautiful', which informed the ethos of CAT's pioneering years, seems anachronistic. Alternative technology (AT), as it was envisaged from the 1970s to the 1990s, is no longer an appropriate narrative and had to be reframed if it was to remain persuasive. While the idea of achieving zero carbon is not founded on a particular philosophical perspective, in the way that Schumacher's ideas underpinned much of the ethos of CAT's early days, it has provided a coherent and plausible alternative narrative.

Some of those who have been involved with CAT over the years perceive that its current narrative, while plausible, is no longer alternative. They suggest that, with the loss of the site community, the introduction of a more hierarchical management structure and the attenuation of the egalitarian ethos, CAT has lost its radical edge. It has, for whatever reasons, made too many compromises and therefore can no longer sufficiently challenge the status quo and the dominant values of society that are the root causes of the environmental crises. For others though, the association with radicalism, in the context of ecological modernisation and discourses of sustainable development, is no longer a viable agenda. Challenge is best achieved working within rather than outside sociopolitical structures. CAT has worked hard to shed its countercultural image and is now perceived as a credible voice, which makes its argument more persuasive. Nonetheless, CAT still represents itself as advocating a radical agenda. On its opening webpage, it invites readers to 'Join the Change', and then indicates that: '*radical action* is needed if we are to avoid dangerous climate breakdown' (CAT 2022b. My emphasis).

CAT, although it is a very different organisation from when it started, can still be considered a practical utopian project. Zero Carbon Britain (ZCB) both imagines a better way of being and proposes plausible ways of achieving a better, carbon-free and sustainable future. The core ideas of ZCB – powering down and powering up – encapsulate the idealism and pragmatism of the CAT project. Powering down, although it includes some practical aspects like improving the energy efficiency for all buildings, is utopian as it calls for lifestyle changes. The latest ZCB Report clearly states: 'The challenge that we face is not only for our technology, but also for our culture' (CAT 2019: 12). The 'powering up' aspect of the ZCB agenda relates more to the practical aspects. This primarily refers to the use of renewable and carbon-neutral means of generating power. Once we have powered down our energy demands, CAT claims that we can supply all our energy needs in the UK using renewable sources. While this seemed implausible in the pioneering days, and even a stretch of the imagination when Bob Todd compiled *An Alternative Energy Strategy* in 1977, with the rapid advances in the hardware of AT, this now seems eminently achievable. The ZCB is adamant that 'we can supply 100% of the UK's powered down demand with renewable and carbon-neutral energy sources without fossil fuels and without nuclear' (CAT 2019: xi).

There is an inherent critique of contemporary ideas of perpetual economic growth and the ethos of consumerism that is regarded as the root cause of the environmental crises in CAT's discourse. 'Conspicuous consumerism now exerts an irresistible pressure, making society reluctant to question the access to the energy supplies that underpins it' (CAT 2019: 13). Implicit in this argument is that we have more material stuff than we actually need. We perceive possessing stuff, such as a car (or cars) as a necessity rather than a want and take for granted lifestyle choices such as having a high meat and dairy diet. We can do with less without sacrificing comfort and convenience. This is not proposing hair shirts and sandals or a return to the dark ages, but is suggesting that we can resist the pressures of consumer society to accumulate ever more stuff.

Changing our lifestyles will not only prevent an environmental catastrophe but is also represented as being beneficial for health and wellbeing.

> Growth in fossil fuelled consumer culture isn't just wrecking the wellbeing of the planet – the tendency to base our identities on money possessions or appearance is also seriously affecting our own health and happiness.
>
> (CAT 2019: 22)

If we seriously address environmental concerns, the argument runs, the overall quality of our life will be improved as a result. Environmentalism is envisaged as not only critical to avert an eco-apocalypse but also as a better way of being not only for the planet but for the health and wellbeing of society and individuals. Socially, decarbonising is linked to climate justice, creating a more equitable future. Individually, decarbonising our lifestyle will improve our physical and psychological health.

One of the most important metaphors in discourses about climate change is that of tipping points. This suggests that there are certain thresholds that, once crossed, produce effects that are irreversible. The IPCC (2014: 128) defines a tipping point as: 'A level of change in system properties beyond which a system reorganizes, often abruptly, and does not return to the initial state even if the drivers of the change are abated'. Paul suggests that there are two tipping points: one that will take humanity and the planet over the edge of irreversible climate change and the other that can bring us back from the brink of environmental cataclysm.

> We are heading to two tipping points. One is a cultural tipping point where we understand what needs to be done and we change, and the other is a climate and biodiversity tipping point where we just tumble into an uninhabitable earth. The issue is which tipping point is triggered first.
>
> (Paul, interview January 2021)

There is a consensus among climate scientists that it is increasingly urgent that we act to mitigate the very worst of environmental catastrophe by ensuring that there is no more than a 1.5°C rise in global warming above pre-industrial levels. Achieving net zero as soon as possible is imperative to achieve this. CAT, as do many others, argue that the current UK Government's target of achieving net zero by 2050 is not sufficient. According to CAT (2019: 12), 'Time is now of the essence, as succeeding slowly is actually failing'.

Ensuring that humanity and the planet are not tipped over into an irreversible climate disaster but is tipped back into an environmental equilibrium requires a critical mass of people, organisations and institutions to act in sustainable ways. Achieving this critical mass requires effective communication and education. CAT has been at the forefront of environmental education for five decades. In the early years of CAT, informing the public was critical as, other than very specific and local disasters such as oil spills, there was little awareness of anthropogenic causes of environmental harm. However, it is no longer the case that people lack information about the human causes of environmental harm. It is also well established that addressing the information deficit is insufficient to persuade people to act. We have too much information of the wrong kind. We are bombarded with an overwhelming picture of a looming eco-apocalypse, which leaves us in a state of eco-anxiety and paralysed with fear. Consequently, we need positive, plausible solutions.

While diagnosis and critique of the present are a necessary starting point, they are not alone sufficient for inspiring a response. Two other ingredients are necessary to actually inspire people to take action: a vision of a better future and some practical steps to achieve that future. In this instance, what is required are some realistic actions that we can take now to ensure a credible and sustainable future. CAT has always had a vision of a better future, from the Village of the Future imagined in the 1970s to the ZCB envisaged today. CAT has always emphasised the practical solutions to the perceived environmental issues of the day as plausible steps to avert the looming eco-apocalypse. What is important is not so much the actual details of the steps proposed by CAT, but that these proposed practical solutions raise the possibility that 'we can do it' – we can avert ecological disaster.

CAT alone is not going to break the paralysis of eco-anxiety, but as an important member of a choir whose combined voices increasingly act to amplify each other, it can no longer be dismissed as quixotic. The more voices that join this chorus, the better chance we have of reversing this trajectory that sees us hurtling towards a tipping point, which if crossed will entail irrevocable environmental devastation. I am hoping that in its own small way, this monograph will add to this chorus and inspire people to imagine a plausible and better future along with the practical steps to achieve sustainability. We need as many practical utopian discourses as possible.

References

Carter, M. (1988). Stasis and Kairos: Principles of Social Construction in Classical Rhetoric. *Rhetoric Review* 7(1), 97–112.

Centre for Alternative Technology (CAT). (2019). *Zero Carbon Britain: Rising to the Climate Emergency*. Machynlleth: CAT. Available at https://cat.org.uk/info-resour ces/zero-carbon-britain/research-reports

Centre for Alternative Technology (CAT). (2022a). Annual Report 2020-21. Available at https://cat.org.uk/annual-report (Accessed 23 February 2022).

Centre for Alternative Technology (CAT). (2022b). https://cat.org.uk (Accessed 27 February 2022).

Intergovernmental Panel on Climate Change (IPCC). (2014). *Annexe II: Glossary.* Available at: www.ipcc.ch/site/assets/uploads/2018/02/AR5_SYR_FINAL_Anne xes.pdf (Accessed 13 February 2022).

Radkau, J. (2014). *The Age of Ecology.* Cambridge, UK: Polity Press.

Rawcliffe, P. (2000). The Role of the Green Movement in Ecological Modernisation: A British Perspective. In S.C. Young (Ed.), *The Emergence of Ecological Modernisation: Integrating the Environment and the Economy.* London: Routledge, 65–86.

Young, S.C. (2000). Introduction. In S.C. Young (Ed.), *The Emergence of Ecological Modernisation: Integrating the Environment and the Economy.* London: Routledge, 1–39.

Appendix
People

This is by no means a complete list of the many people who have been involved in CAT over the years. Nor is it a complete list of people who were interviewed for the oral history project. The list is only of the people that I refer to in the book. The numbers in the brackets refer to the catalogue number of the recordings held at the National Library of Wales in Aberystwyth.

Billy Aiken: Digital Marketing and Web Design from 2021.

Paul Allen: Came as an electronic engineer. He went to work for Dulas Engineering before returning to CAT as Media Officer. He is now the Outreach Coordinator for ZCB. (Recording 1-1-1-1).

John and Audrey Beaumont: The original landowners and landlords before CAT bought the site in 2004. (Recording 8-1-3).

Don Bennett: Moved to Machynlleth from South Wales to set up a pottery. Briefly worked at CAT as the caretaker for the WISE building. (Recording 3-5-1).

Pat Borer: An architect. He worked at CAT between 1975 and 1990. He designed both the AtEIC Building and WISE with David Lea. Now lectures for the GSE. (Recordings 6-1-1; 6-1-2 and 6-1-3).

Godfrey Boyle: Editor of *Undercurrents* and joint editor with Peter Harper of *Radical Technology*. He established the Alternative Technology group at the Open University. (Recording 5-6-1).

Diana Brass: One of the original members of CAT. She accompanied Gerard Morgan-Grenville on his exploratory trip to the USA. Diana stayed on site while Gerard Morgan-Grenville mostly worked in the background.

Anna Bullen: Joined CAT as the ZCB Innovation Lab Manager in 2020.

Rob Bullen: Visitor Marketing and Business Development Manager.

Sabrina Cantor: Came to CAT in 1982 and worked at CAT for about 20 years. She mostly worked in bookshop and the Mail Order Department. (Recording 6-5-1).

Sally Carr: Came as a volunteer in 1986, returned to run the volunteer programme in 2001 and was then appointed to the Operations Team in 2010. She now sits on the Board of Trustees.

John Challen: Head of Eco Centre.

Rick Dance: Came as a volunteer in 1983. He then went to work at the Urban Centre for Appropriate Technology before returning to CAT to work as an administrator in 1987. He was appointed as Company Secretary when CAT launched the share issue. He left CAT in 2015. He was a member of the site community for several years. (Recordings 6-8-1 and 6-8-2).

Nigel Dudley: A very early member of CAT. He worked in the Quarry Café & Shop and was CAT's first Education Officer. (Recording 5-8-1).

Jonathon Gross: Ran the Quarry Shop for about seven years after the shop acquired a separate premise from the café. (Recording 4-6-1).

Cindy Harris: Builder. She came to CAT as a volunteer and stayed for 17 years. She was responsible for managing the build of the Top Station for the cliff railway and the Autonomous Environmental Information Centre AtEIC. She lived on-site for most of this time. (Recording 2-3-1-1).

Peter Harper: Worked at CAT between 1983 and 2010 in various roles including volunteer coordinator, site management and biology department. He coined the term 'alternative technology'. He was a member of the site community for the early part of his time at CAT. (Recordings 2-1-3-1; 2-1-3-3 and 2-1-3-4).

Phil Horton: Came to work in the Information Department in 1988. He was appointed project manager for WISE. He left after the WISE project to work for Dulas in 2010. (Recordings 3-7-1 and 3-7-2).

Roderick James: Architect and CAT's first Director 1975–1980. (Recording 1-1-3-1).

Edward Jones: Local farmer and friend of CAT. He helped CAT on a number of major projects and set up one of the UK's first wind farms. (Recordings 7-2-1 and 7-2-2).

Hew Jones: Local farmer and parish councillor. (Recording 7-1-1).

Roger Kelly: CAT's third Director 1989–1998 (Recording 1-1-5-1).

Tim Kirby: Site engineer 1982 to mid-1990s. He lived on site for a number of years. He went on to co-found EcoGen, a supplier of renewable energy.

David Lea: Architect who worked with Pat Borer designing the AtEIC Building and WISE. (Recording 2-3-2-1).

Jeremy Light: Head of Biology Department. He was responsible for installing CAT's first reed bed system for sewage treatment. He worked at CAT between 1976 and 1993. (Recordings 2-1-4-1 and 2-1-4-2).

Rachel Lilley: Development Coordinator and Fund Raiser. She worked at CAT for several years from1994. (Recordings 3-6-1 and 3-6-2).

Annie Lowmass: Joined CAT in 1981. She ran the Quarry Shop and Café for a number of years. She was part of the site community. She then took a short break and returned to run the retail and mail order side of the organisation. (Recordings 4-3-1 and 4-3-2).

Kelvin Mason: Worked at CAT for a couple of years in the late 1970s/early 1980s. He returned in about 2009 as a lecturer for the GSE. (Recording 5-5-1).

Mark Mathews: He and his wife Mary were appointed by Morgan-Grenville as *de facto* site managers in 1974. They took over when Tony Williams the very first site manager left. (Recordings 1-1-2-1 and 1-1-2-2).

Gerard Morgan-Grenville: The founder of CAT.

Clive Newman: Joined CAT in 1986. He was site engineer and member of the site community. (Recording 2-2-2-1).

Caroline Oakley: Head of publications 1999–2008. (Recording 6-3-1).

Graham Preston: Graphic designer. He worked at CAT for almost 20 years. (Recording 6-2-2).

Pete Raine: CAT's second Director and member of site community 1980–1986. He worked for FoE before joining CAT. (Recording 1-1-4-1).

Joan Randle: Came to work at CAT in 1982 as Educational Officer, in a job share with her husband Damian. She later developed CAT's course provision and was instrumental in the development of Postgraduate courses and the GSE. (Recordings 5-4-1 and 5-4-2).

Delyth Rees: Local resident and was a trustee for a number of years. (Recording 1-2-1-1 in Welsh).

Andy Rowland: Ran CAT's bookshop and then became site manager. Joined in 1984 and worked at CAT for 14 years. He went on to manage Ecodyfi, a local sustainability organisation. (Recording 4-1-1).

Allan Shepherd: Came to CAT as a long-term volunteer in 1994. He then got a post in CAT's publishing department. Shepherd organised the Oral History Project. He wrote *The Little Book of Slugs* and *Voices From a Disused Slate Quarry*. (Recordings 6-6-1 and 6-6-2).

Amanda Smith: Joined CAT in 2018 to work in the Engagement Team. She is now the training manager for ZCB.

Ruth Stevenson: Senior Lecturer at the GSE.

Richard St George: A very early volunteer. He went on to become the Director of the Schumacher Society.

Mick Taylor: Chair of the Board of Trustees for 10 years until March 2022.

Bob Todd: CAT's first Technical Director. He was responsible for the 1977 publication *An Alternative Energy Strategy for the United Kingdom.* He went on to start Dulas Engineering and later Aber Instruments. (Recording 2-2-5-1).

Peter Tyldesley: CEO 2019 to 2022.

Liz Todd: Bob Todd's wife. She was appointed as Machynlleth's town librarian. (Recordings 1-2-4-1 and 1-2-4-2).

John Urry: Came to CAT in 1982 for a university project and stayed. He works in graphics and for most of his time has lived on site. (Recording 6-9-2).

Adrian Watson: Head of the GSE.

Petra Weinmann: Display Gardener.

Merri Wells: A local of Machynlleth who worked at the Quarry Shop when it first opened. She is now an internationally renowned ceramics sculptor. (Recordings 4-2-1 and 4-2-2).

Jill Whitehead: An early volunteer who was then employed to develop and run the courses programme. (Recordings 1-2-5-1 and 1-2-5-2).

Index

For Product Safety Concerns and Information please contact our EU
representative GPSR@taylorandfrancis.com
Taylor & Francis Verlag GmbH, Kaufingerstraße 24, 80331 München, Germany

www.ingramcontent.com/pod-product-compliance
Lightning Source LLC
Chambersburg PA
CBHW060301220326
41598CB00027B/4190

* 9 7 8 1 0 3 2 0 7 5 6 3 1 *